Generis

PUBLISHING

AF722412

SIMPLE AND COMPOUND MECHANICAL TESTS ON WIND TURBINE BLADES

Ciprian-Ionuț Morăraș
Viorel Goanță

Title: **SIMPLE AND COMPOUND MECHANICAL TESTS ON WIND TURBINE BLADES**

ISBN: 979-8-89966-337-6

Author: Ciprian-Ionuţ Morăraş, Viorel Goanţă

Cover image: Dr. Ciprian Morăraş and Prof. Dr. Viorel Goanţă

Publisher: Generis Publishing
Online orders: www.generis-publishing.com
Contact email: info@generis-publishing.com

Ciprian-Ionuț Morăraș

Viorel Goanță

SIMPLE AND COMPOUND MECHANICAL TESTS ON WIND TURBINE BLADES

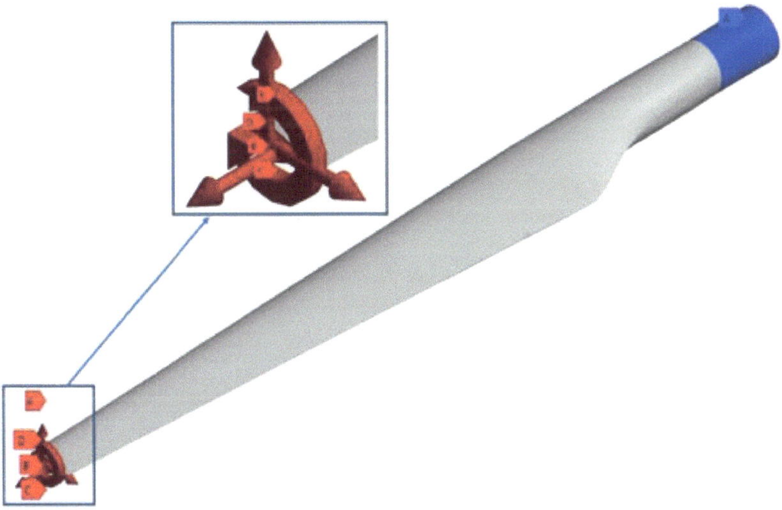

Contents

Preface **9**
Summary **11**
CHAPTER 1. STATE-OF-THE-ART IN WIND TURBINE BLADE ANALYSIS, TESTING AND MONITORING **13**
1.1 Blade loads during operation *13*
1.2 Finite element analysis applied to wind turbine blades *15*
1.3 Blade testing *22*
1.4 Blade monitoring during operation *45*
1.5. General conclusions to Chapter 1 .... *52*
CHAPTER 2. FINITE ELEMENT ANALYSIS OF SMALL-SCALE WIND TURBINE BLADES **53**
2.1 The scope of finite element analysis *53*
2.2 Wind turbine blade design and construction *53*
2.3 Pre-processing the wind turbine blade model *55*
2.4 Case 1: Finite element analysis for bending: F= 1000 N *56*
2.5 Case 2: Finite element analysis at clockwise torsion Mt= −170660.646 N·mm *64*
2.6 Case 3: Finite element analysis at anticlockwise torsion M_t*= 170660.646 N·mm* *68*
2.7 Case 4: Finite element analysis under compound load (bending + clockwise torsion) F=1000N + Mt= −170660.646 N·mm *72*
2.8 Case 5: Finite element analysis under compound load (Bending + Anticlockwise torsion F=-1000N + Mt= 170660.646 N·mm) *78*
2.9. General conclusions to Chapter 2 .... *86*
CHAPTER 3. TESTING A SMALL-SCALE BLADE UNDER COMPOUND LOAD CONDITIONS **87**
3.1 The purpose of testing the wind turbine blade at scale *87*
3.2 The experimental set-up and sensors used *87*
3.3 Electrical resistivity tensometry *93*
3.4 Determination of stresses and strains on wind turbine blades *96*
3.5 Calibration with weights under bending and torsional stress *98*
3.6 Experimental results obtained on rectangular transducers *102*
3.7. Experimental results obtained on delta-type strain gauge chains *129*
3.7.1 Blade bending load *129*
3.7.2 Blade Torsion Loading *159*
3.7.2.1 Clockwise Torsion Loading *159*

3.7.2.2 Anticlockwise torsion loading. *164*
3.7.3 Compound load of bending and torsion of the wind turbine blade *169*
3.7.3.1 Bending with clockwise torsion loading *170*
3.7.3.2 Bending with anticlockwise torsion loading *178*
3.8 Validation of experimental results with FEM *187*
3.9 General conclusions to Chapter 3 .... *194*
References **195**

Preface

In the context of the increasing importance of renewable energy, wind turbine blades are critical components, whose durability and performance directly influence the efficiency and safety of the entire system. Composite materials, especially those based on glass fiber (GFRP), are widely used due to their advantageous strength-to-weight ratio. However, the complexity of the mechanical stresses to which blades are subjected during operation – including aerodynamic, gravitational, centrifugal and dynamic forces – requires the development of rigorous testing and monitoring methods.

The paper presents the investigation of the structural behavior of a 1:10 scale GFRP wind turbine blade, through an integrated approach that combines numerical analysis using the finite element method (FEM) with experimental testing under controlled laboratory conditions. It aims to identify the distribution of stresses and strains under simple and compound loads, as well as to validate the numerical models through direct measurements using resistive electrical strain gauges. The results obtained contribute to the optimization of testing and monitoring methods used in the industry.

The research hypothesis states that numerical simulations performed using FEM, calibrated with precise experimental data, can provide correlated and relevant results, facilitating the identification of critical areas of the blade and the optimal positioning of monitoring sensors.

To achieve this goal, the paper proposes the following specific objectives: developing a three-dimensional model of the blade in ANSYS Academic R17.2, defining the specific material characteristics of GFRP, simulating five simple and compound loading cases to evaluate the stress distribution, determining the optimal positions for the placement of strain gauge sensors, performing experimental tests in ten loading and unloading scenarios, correlating the numerical results with the experimental ones and determining the principal strains and stresses according to the indirect method recommended by Micro-Measurements Vishay. This approach provides a solid basis for developing effective methods for monitoring and increasing the durability of these essential components of wind systems.

Summary

The book entitled ***Simple and Compound Mechanical Tests on Wind Turbine Blades*** is the result of extensive research dedicated to simple and compound stress testing of a small-scale wind turbine blade made of GFRP (glass fiber reinforced plastic). This work is the sequel of a previous publication entitled ***Testing of Composite Materials Used in Wind Turbine Blades***.

Chapter 1 addresses the current state of research and technologies related to the testing and monitoring of wind turbine blades. During operation, wind turbine blades are subjected to a variety of loads. These include aerodynamic forces generated by interaction with the airflow, gravitational loads due to their own weight, centrifugal forces resulting from rotor rotation, and dynamic loads caused by gusts, turbulence, or variations in wind conditions. All these loads can induce vibrations, stresses, and deformations in the structure, affecting the performance and durability of the blades. In order to understand the structural behavior of the blades under these loads, the finite element analysis (FEA) method is used.

Chapter 2 presents the finite element analysis (FEA) of a 1:10 scale wind turbine blade, performed using ANSYS Academic R17.2. The blade's material properties were specified for an E-glass composite. The characteristics of the experimentally obtained material were implemented into the ANSYS material library. It was estimated that the maximum loading the blade model could be subjected to includes a force of F = 1000 N (applied at the blade tip) and a torque of Mt = 170.660 N·mm (also applied at the tip). FEA simulations were conducted for five different loading cases: bending F=1000N; clockwise torsion M_t= −170660.646 N·mm; anticlockwise torsion M_t= 170660.646 N·mm; bending + clockwise torsion F=−1000 N + M_t= −170660.646 N·mm; bending + anticlockwise torsion F= 1000N + M_t= 170660.646 N·mm. The FEA results, particularly the distribution of maximum von Mises equivalent stress and maximum principal stress, were used to determine the optimal placement of four strain gauge rosettes and four strain gauge chains on the blade surface.

Chapter 3 is dedicated to experimental testing of the wind turbine blade under simple (bending and torsion) and combined (bending with torsion) loading conditions. Ten loading and unloading cases were studied: simple bending –

loading, simple bending – unloading, clockwise torsion – loading, clockwise torsion – unloading, anticlockwise torsion – loading, anticlockwise torsion – unloading, combined clockwise torsion and bending – loading, combined clockwise torsion and bending – unloading, combined anticlockwise torsion and bending – loading, and combined anticlockwise torsion and bending – unloading.

To determine the strain values, the resistive electrical strain gauge method was used. Four rectangular strain gauges were applied in critical zones of the blade, and twelve specific strain measurements were acquired using a Vishay P3 strain gauge bridge. Additionally, four strain gauge chains with delta-type transducers provided a number of 60 strain measurements using the QUANTUM HBM MX1615B data acquisition system. Using the indirect method recommended by Micro-Measurements Vishay, the principal strains and their directions, as well as the principal stresses and their orientations, were determined for each rosette.

The authors would like to thank Prof. Dr. Eng. Paul-Doru Bârsănescu, Faculty of Mechanical Engineering, Department of Strength of Materials, "Gheorghe Asachi" Technical University of Iași, for their collaboration, support and constructive assistance. The authors also thank their colleagues, Lecturer Dr. Eng. Dorin Husaru and Engineer Axinte Toader.

Authors,
04.12.2025

CHAPTER 1. STATE-OF-THE-ART IN WIND TURBINE BLADE ANALYSIS, TESTING AND MONITORING

1.1 Blade loads during operation

During operation, the wind turbine blade is subjected to the following loads: bending in two planes, axial loads and torsion (Morăraş et al., 2023; Morăraş et al., 2022; Ghafafian et al., 2021), as shown in Figure 1.1. The axial loads are produced by the blade's own weight (to which can be added the weight of the dust deposited on the blade) and centrifugal forces. These two forces act in the same direction when the blade is upright below the hub and in opposite directions when the blade is upright above the hub. Thus, a complex state of tension is created in the blade. The highest stresses are introduced at blade bending. The blade loads are variable during operation.

Due to fatigue, cracks continue to progressively enlarge and may lead to blade failure. Blade sheath buckling is a complex phenomenon. Laboratory tests (including fatigue) cannot predict all loads and environmental conditions to which the blade is subjected in service.

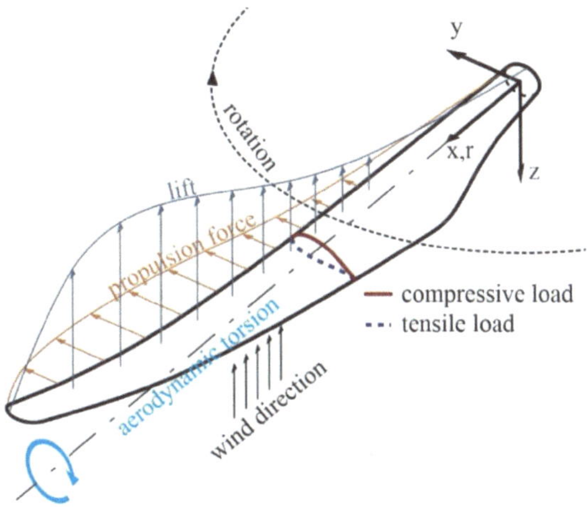

Fig. 1.1 Blade loads during operation (centrifugal force not shown) (Ghafafian et al., 2021)

Composite materials with a polymetric matrix are hydrophilic. The absorption of water into the blades has multiple effects:

- increasing the weight of the blades, thus increasing their loads;
- producing imbalances, leading to vibrations;
- moisture absorption in composite materials leads to a decrease in their mechanical and elastic characteristics. The effects of moisture on the mechanical properties of fiber reinforced materials (FRP) are summarized in Table 1.3 (Katsaprakakis et al., 2021; Morăraș et al., 2020).

Table 1.1 The influence of moisture on some characteristics of composite materials with polymer matrix (Katsaprakakis et al., 2021)

Composite material	Maximum humidity Absorption (%)	Stiffness variation (%)	Fatigue endurance Number of cycles
Glass - polyester	4	-10	-35×10^{-3}- -10×10^{7}
Glass – epoxy	2	-10	-20×10^{-3}- -5×10^{7}
Carbon – polyester	1,5	1	
Carbon – epoxy	1,5	1	±0
Glass-carbon – epoxy	<2	±0	

A decrease of up to 36% in the interlaminar strength of the glass/epoxy composite material used in making blades due to moisture has been reported by Rocha et al., as shown in Figure 1.2a. The influence of moisture on GFRP material is shown in Figure 1.2 b (Rocha et al., 2017).

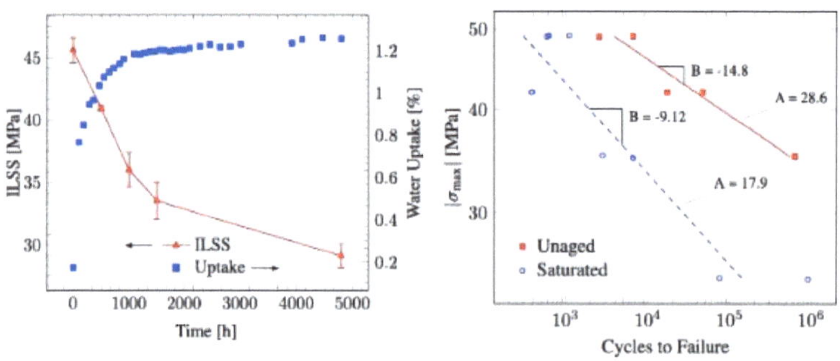

a) b)

Fig. 1.2 Influence of moisture on the interlaminar shear strength (a) and fatigue behavior of GFRP (b) (Rocha et al., 2017)

Frost deposition on the blades also causes rotor imbalance and vibrations. In addition, frost deposition modifies the aerodynamic profile of the blade (Figure 1.3), leading to a drastic reduction in wind turbine performance, possibly requiring its shutdown.

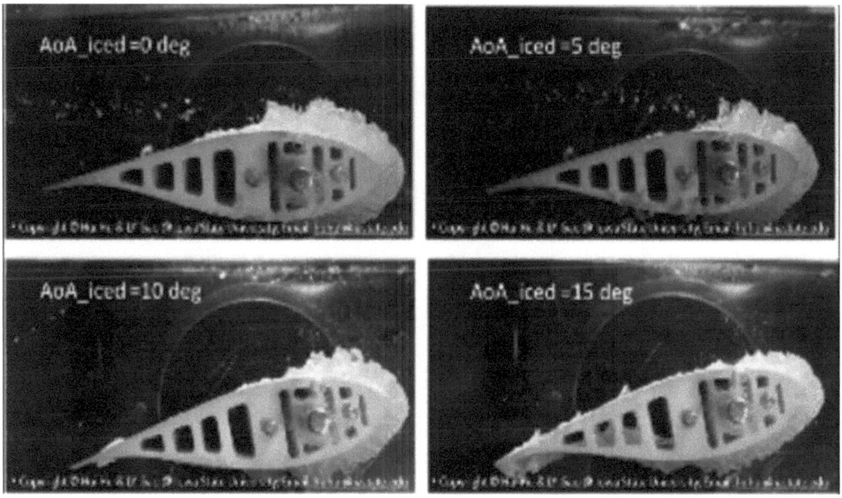

Fig. 1.3 Ice deposition, at different temperatures, after 10 minutes of rotor operation in the wind tunnel (https://www.canarymedia.com/articles/enn/iowa-state-researchers-seek-solution-for-icy-wind-turbine-blades)

1.2 Finite element analysis applied to wind turbine blades

The geometry and loading of wind turbine blades are complex and, for this reason, the analytical approach can only provide approximate results (Tarfaoui et.al., 2019; Hasen, 2008).

For this reason, blades are designed using numerical simulation methods, especially the finite element method. Finite element analysis (FEA) thus becomes an indispensable tool in studying the state of loads, deformations, displacements and stresses in wind turbine blades.

An important advantage of FEA is that this method can simulate both real loads during operation and concentrated forces and moments applied during blade testing.

The following main loads act on the wind turbine blade:

- pressures from the blade/airflow interaction;
- mass forces (gravitational and centrifugal).

The turbine blade must be designed for worst-case conditions, when the loads are maximum. The determination of the pressure loads can be achieved using experimental methods and/or numerical methods (AEF), considering the fluid/solid interaction.

In of the Liu et al. (2022), the ABAQUS program was used to simulate a blade of length 59.5 m. The material for the cover and spar was a glass/epoxy composite with the layer arrangement at $[0^o/45^o/90^o/ -45^\circ]_{12}$ + $[0^o/45^o]$ and $[0^o/45^o/90^o/ - 45^o]_5$ respectively. The density of the composite material was 1200 km/m^3. The blade was discretized into 7700 finite elements, sufficient to ensure the required accuracy, as shown in Figure 1.4.

We started from the fluid/solid interaction to determine the stress states, deformations and displacements in the blade for different wind speeds. Wind shear effect, tower influence and predicted blade lifetimes were considered using different blade structure failure models. The maximum stresses appear in the blade sheathing as follows, as shown in Figure 1.5:

- In the reinforcement layer oriented at 45° from the hub area;
- In the airfoil about 1/3 of the length from the blade tip.

The adhesive bond between the spar and the analyzed blade sheathing starts to fail at a wind speed of 15 m/s and a tangential stress of 5 MPa. At wind speeds above 25 m/s and a tangential stress of 30 MPa, the adhesive joint fails completely. Compressive fiber yielding occurs in the hub area and it is more pronounced near the leading edge. The tensile fiber yielding occurs, in particular, on the spar. Compressive matrix failure occurs mainly on the outer layers of the sheathing.

Tensile matrix failure occurs mainly in the area of the hub and the joint between the spar and the blade skin. For low wind speeds, the analytical method gives satisfactory results, but for high wind speeds, which are determined from the solid/fluid interaction, it can accurately predict the state of stresses, strains and displacements and, based on these, the mode of blade failure.

In of the Liu et al. (2022), the failure mechanisms of a wind turbine blade reinforced with type I shear webs considering the fluid/solid interaction, as shown in Figure 1.4, were studied using FEA. It predicted the progressive growth of inter- and intralaminar cracks as well as the failure of the adhesive layer between the stringer and the blade shroud, leading to the following conclusions:

- fiber and matrix breaks, as well as delaminations, occur at wind speeds above 20m/s and tangential stresses above 30 MPa;
- adhesive bond failure occurs at wind speeds above 20 m/s and tangential stresses above 5 MPa.

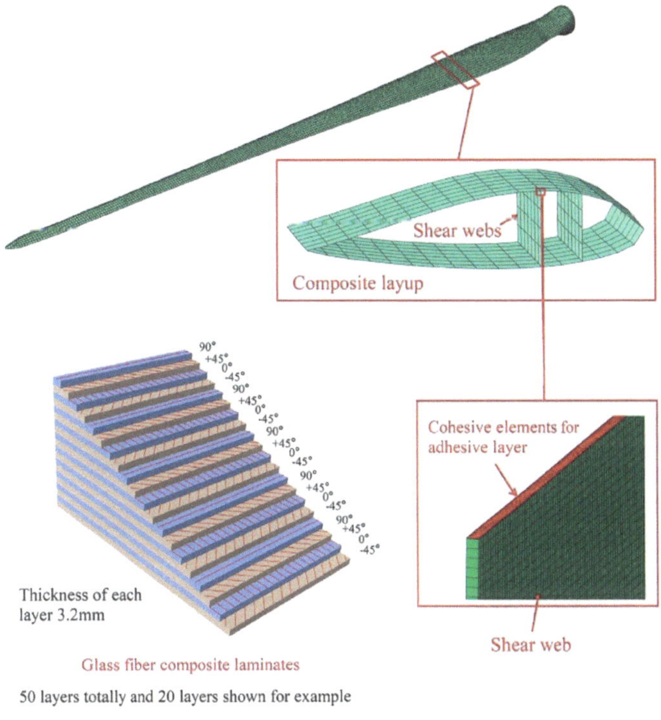

***Fig. 1.4 Mesh model for composite blade with layups* (Liu et al., 2022)**

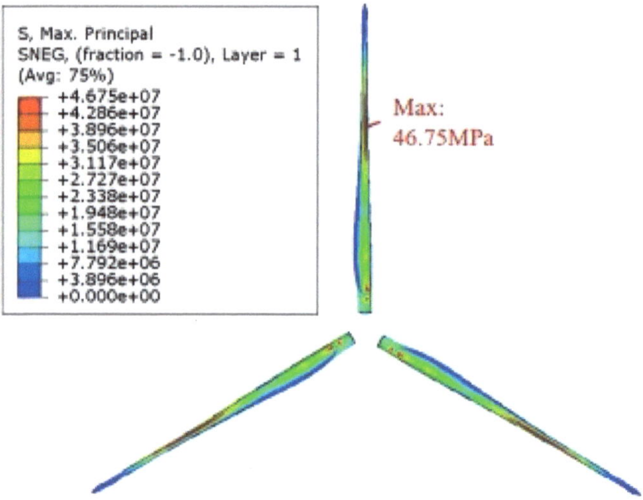

Fig. 1.5 Numerical results for blade under the rated wind velocity 10 m/s: distributions of the maximum principal stress (Liu et al.,2022)

In of the Tavares et al. (2022), the torsional loading of a blade was studied and modeled with the help of FEM, with the loads occurring in service. The finite element analysis was carried out in two variants: with discretization having solid finite elements and with shell finite elements, respectively. It was found that the stiffness prediction can differ by up to 30%. In order to understand this difference, the authors studied a tube subjected to torsion, a problem for which analytical solutions also exist. In this simple case, the improved solid finite element model predicted a 5% lower torsional stiffness than the shell finite element modeling, with more accurate solutions given by the shell finite element analysis.

In of the Wang et al. (2023), FEM and torsion testing of a blade segment are presented. The testing was carried out using a stand and the specific deformations were measured using strain gauge mounted on the blade segment. FEM was performed using the commercial software ABAQUS 16.0 with the ABAQUS/ Standard solver. The discretization was done with shell finite elements. Modeling was done on the blade segment embedded at one end and loaded with a torsional moment at the free end. The maximum von Mises stress of 24.5 MPa was localized in the sandwich composite in the trailing edge area in the vicinity of the embedment, as shown in Figure 1.6.

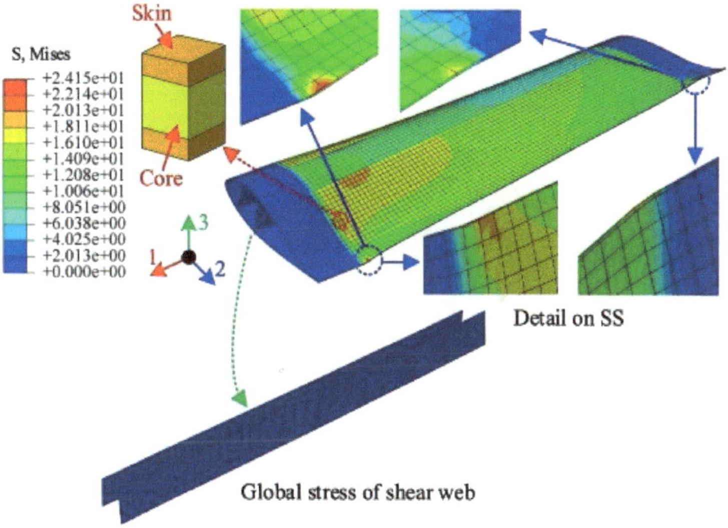

Fig. 1.6 Global stress results of the blade segment through FEM (Wang et al.,2023)

In of the Xiao et al. (2014), a numerical simulation of the static bending test of a 52.3 m long commercial blade is presented. The blade was embedded at the hub and forces were applied to it in four sections. The sections and forces were chosen to best simulate the stresses, strains and displacements during turbine operation, as in Figure 1.7. 47,000 "SR4" shell elements and 33,000 "C3D8RC3" solid elements were used. The main goal of this paper is to predict the blade failure modes using numerical modeling. The following conclusions were drawn from the simulations:

- major failures are localized in the transition area between the aerodynamic part and the hub;
- delaminations were observed in the transition area and shrinkage of the section, delaminations between the core and the skin of the sandwich composite as well as failures of the spar;
- the appearance of delaminations and spar failure in the transition zone were identified as primary causes of blade failure;
- local buckling contributes to blade failure through the occurrence of deformations and displacements in the direction normal to the blade surface;

- variable loading has been simulated where cracks are initiated in the transition zone;
- The study revealed the influence of spatial stress states on the initiation of blade failure.

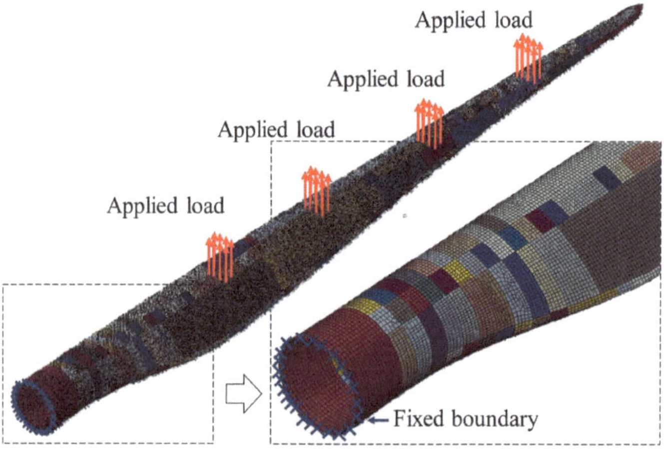

Fig. 1.7 The global shell element model with applied loads (Xiao et at.,2014)

In of the Finnegan et al. (2021) analyzed a scaled-down blade model (13 m in length) to simulate the testing of a 10 MW turbine blade with a rotor diameter of 190m, subjected to static bending. The blade model was embedded to the hub and loaded with applied forces in three sections. The blade model was loaded in two directions: intrados - extrados and leading edge - trailing edge, respectively, as in Table 1.2. The force application distances are measured from the blade hub.

Table 1.2 Loads applied to the blade model subjected to bending (Finnegan et al., 2021)

Blade position	Force (KN) at a distance of 5.75 m	Force (KN) at a distance of 8.5 m	Force (KN) at a distance of 10.95 m
Horizontal (loading in the intrados-extrados direction)	5.6	11.9	2.3
Vertical (loading in the leading edge-trailing edge direction)	6.3	4.8	2.4

AEF was performed using three simulation programs: ANSYS, ABAQUS and CalculiX. The results obtained by FEM and experimentally were compared with each other. Figures 1.8 and 1.9 present a comparison between the variation of specific elongations obtained along the blade by the three numerical simulation models and by means of strain gauges.

In Figure 1.8, it is observed that the specific deformations are positive on the extrados and negative on the intrados of the blade. The three FEM models generally underestimate the values obtained experimentally. The Ansys and Abaqus modeling provide similar results, which are in good agreement with the experimental ones in the regions 0.75 – 2 m and 7 – 10 m respectively. In the region 3 – 7 m, a reasonable agreement with the experimental results is obtained.

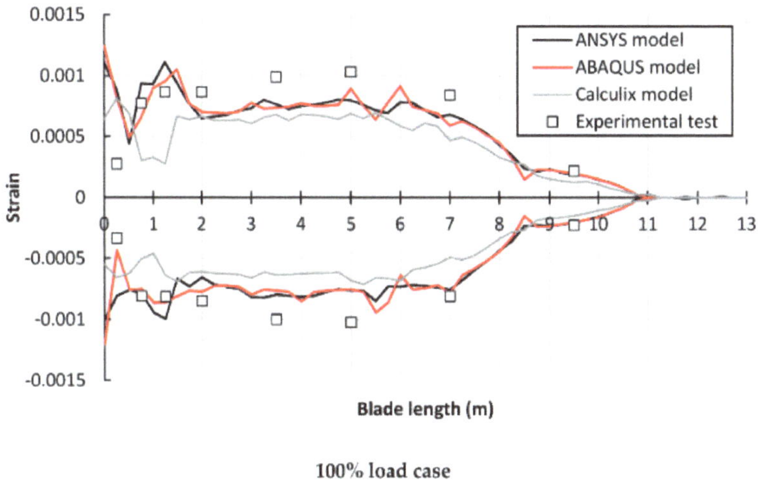

Fig. 1.8 Comparison of results for 100% loading of the blade placed in a horizontal position

Figure 1.9 shows that the specific deformations are positive on the leading edge and negative on the trailing edge. It can be seen that, also in this loading case, the three AEF models generally underestimate the experimentally obtained values. The CalculiX model provides results that are in good agreement with those obtained experimentally in the leading-edge region. On the trailing edge, the results obtained by Ansys and CalculiX agree satisfactorily with the experimental results in the hub area and underestimate them in the rest of the blade. The AEF performed with Abaqus shows some larger jumps and errors than the experimental results.

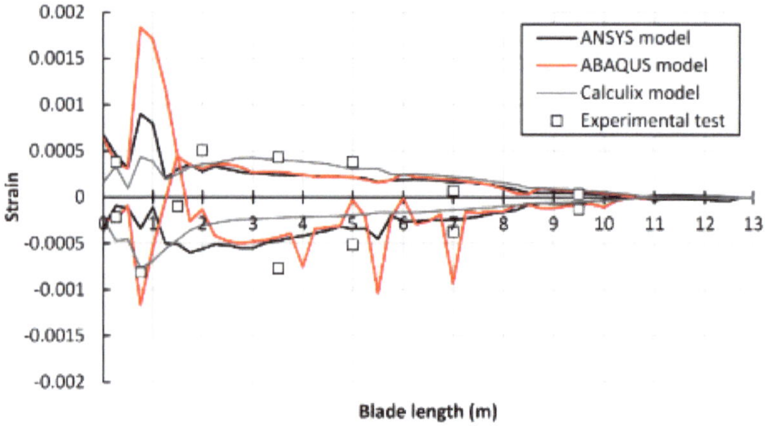

100% load case

Fig. 1.9 Comparison of results for 100% loading of the blade placed in a vertical position

During the tests, the deflections in the force application sections (4 m, 7.5 m, 8 m and 13 m from the hub) were also measured. The estimates made with the three AEF models generally agree well with the experimental determinations:

- ANSYS provides results that differ by 1.1% from the experimental ones;
- The results provided by CalculiX differ by -3.3% from the experimental ones;
- ABAQUS gives results that differ by -11.5% from the experimental ones.

1.3 Blade testing

The determination of the elastic and mechanical properties of the materials from which the blades are made can be carried out on specimens or sub-components made by the same technologies as those used in the manufacture of the blades.

Testing blades, especially large ones, under all these simultaneously acting loads is very difficult, due to the size and geometry of the blades as well as the complex loads to which they are subjected during operation. For this reason, full-scale blade tests are often performed only at bending (static or variable loading), with the blade placed horizontally and then vertically, and sometimes also at compound loading, usually without applying all loads simultaneously. The main purpose of full-size blade testing is to validate the blade design.

Static tests can be destructive and non-destructive. Their purpose is to test the ability of the blades to withstand the extreme loading, storm or transient stresses, which may occur during rotation of the nacelle. Another purpose of the static tests is to determine the limit of strength of the blades.

The blades are tested in static and variable load tests. According to (International Electrotechnical Commission. Wind, 2001), the fundamental purpose of static testing is to guarantee, for a type of blade, a reasonable level of safety and that stress limit states will not be reached during operation. Static tests of full-size blades are essential for understanding blade failure mechanisms, verifying Non-Destructive Testing (NDT) and Structural Health Monitoring (SHM) techniques (Zhou et al., 2014).

Load tests are carried out on the blades at nominal forces. In the certification process, the blades shall withstand these loads without damage. In strength tests, the blades shall be subjected to static loading until failure. These tests allow a verification of the strength of the blade and the failure zones and mechanisms and provide useful data for the improvement of design models (Malhotra et al., 2012).

There are few dedicated machines for testing wind turbine blades. One of these, produced by MTS, is shown in Figure 1.10. Forces are applied at several points with linear winches and actuators, the loads being coordinated with each other. Static loading of the blades is usually done with electrically driven winches. With this system, applied forces and displacements are easily controlled. Hydraulic actuators are also used for the same purpose, but they are expensive when large displacements are required (https://www.mts.com/en/products/energy/wind-turbine-blade-static-systems). However, many manufacturers make their own stands for testing shovels.

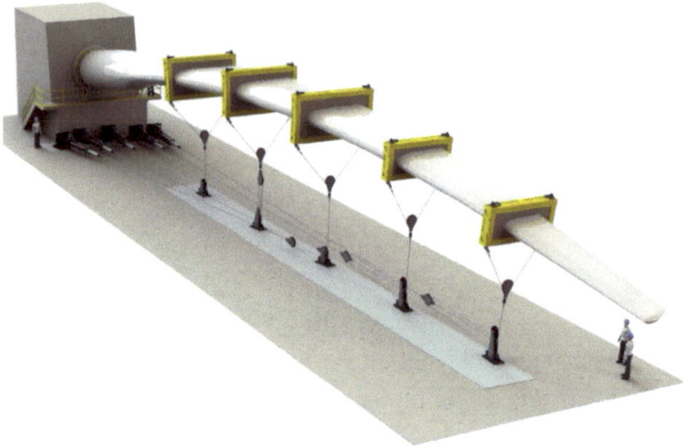

a)

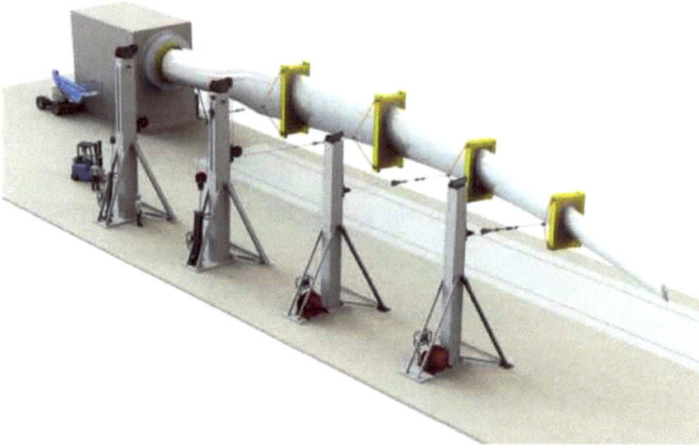

b)

Fig. 1.10 MTS machine for the static bending test of blades: a) Application of forces vertically (in the extrados-intrados direction), b) Application of forces horizontally (in the leading edge - trailing edge)

The European Commission initiated the EWTTPD project in 1996 to develop a single standard for the testing and certification of wind turbine blades and the validation of a new blade design. The project tested 25 blades, 25 m in length, made of fiberglass reinforced polystyrene. The blades were statically loaded to 75% and 110% of the maximum design load respectively. During the tests, the blades were monitored by NDT means. On each blade, 36 strain gauges were mounted. The bending loads were applied in the extrados - intrados and leading edge - trailing edge direction, respectively. A linear behavior of the specific deformations was observed, indicating that no structural failure or loss of stability (buckling) occurred (Bulder, 2002).

The usual static tests are performed as follows:

- Bending, with the blade embedded in the hub and loaded with a number of vertical forces (up to 4), so placed as to simulate as closely as possible the conditions during operation. The blade is loaded with forces acting in the intrados - extrados direction (horizontal position) and in the leading edge-trailing edge direction (vertical position) respectively;
- Torsion, generally carried out on portions of the blade;
- Compound loading (double bending, torsional bending, etc.);
- Blade testing can also be done on scaled models, respecting similarity criteria. Testing on models has the advantage of lower costs and better accuracy of measurements due to the fact that the tests are done in the laboratory. However, tests at varying loads cannot be done on models because cracks start from material defects and the probability of larger micro-defects is higher with increasing material volume.

E-epoxy glass turbine blades for 750 kW wind turbines have been tested at static loads simulating aerodynamic loads and have been shown to maintain structural integrity. The experimental results for blade mass, blade center of gravity position, blade tip arrow and blade natural frequency are in good agreement with those predicted at design, the differences being about 4% (Kong et al., 2005). The Danish manufacturer Risø tested 25m turbine blades made of fiberglass-epoxy (prepreg composite) in order to study the influence of model tests, to classify blade failure modes, to understand the failure mode under compressive stresses and the failure mode of adhesive joints under composite stresses. Seven failure modes produced by static loads have been identified (Sørensen et al., 2004; Debel, 2004; Jørgensen et al., 2004):

- Formation and propagation of cracks in the adhesive layer between the spar and the blade skin;
- Formation and propagation of cracks in the adhesive layer joining the two sides of the blade in the area of the leading or trailing edge;
- Formation and propagation of cracks in the adhesive layer at the core-shell interface of sandwich composites;
- The formation and propagation of interlaminar cracks (delaminations) in the shell and spar (in tension or compression with buckling);

- Rupture of fibers in the shell and spar by tension and failure of the laminate by compression;
- Shell buckling. The formation and increase of cracks in the adhesive joint between the shell and the spar due to buckling;
- Formation and propagation of cracks in the protective gel, gel peeling from the coating.

In of the Chen et al. (2011), tested two types of blades with the dimensions of 1m and 2.5m, the shell was made of epoxy glass and the core of PVC foam. The number of laminations was 8 in the vicinity of the hub and 4 at the blade tip. The inside of the hub was made of aluminum alloy. FEA of the blade was done using the Ansys program, using *shell* finite elements for the shell and *solid* finite elements for the core. Failure of the first layer was predicted using the limit state criterion. In place of cracking, the material characteristics were modified according to the failure diagrams, modifying the stiffness matrices accordingly. The blade was loaded in increments with weights, using a lever and cable mechanism (Whiffle type), until blade failure. At distances 180, 426, 714, 960 mm from the hub, the forces 0.07 F, 0.41 F, 0.31 F and 0.21 F were applied, as shown in Figure 1.11.

Fig. 1.11 Loading mechanism of the blade (static loading in steps, using weights added in buckets) (Chen et al., 2011)

The blade failure occurred at a force of F=1236 N. The blade failed in the hub area at the joint between the aluminum hub core and the composite shell, as shown in Figure 1.12 (Chen et al., 2011).

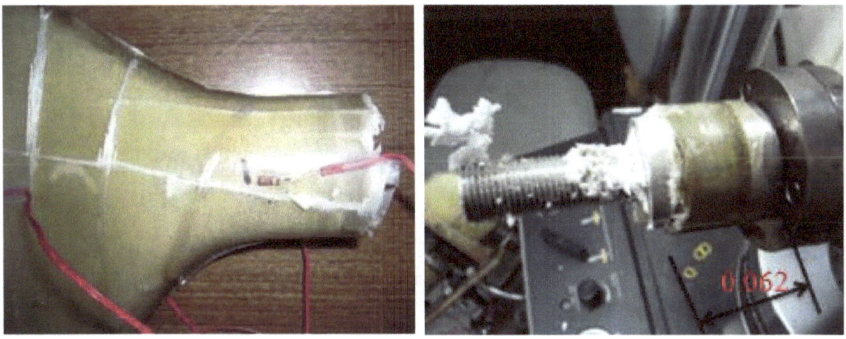

Fig. 1.12 Blade failure with a length of 1m (Chen et al., 2011)

In of the Jinshui et al. (2013), studied the failure mode of a 1:1 scale, 40 m long, fiberglass epoxy-coated, 1:1 scale blade designed for 15 kW wind turbines. The horizontally placed blade was fixed in cantilever and loaded to bending with forces applied in steps in four sections, as shown in Figure 1.13. The failure of the blade with a length of 1m.

Fig. 1.13 The bending load of a 40m long blade (Jinshui et al., 2013)

The blade failure occurred at a force of 160% of the maximum design force and a deflection of 11m at the blade tip. Figure 1.14 shows the exact area and mode of failure of the blade.

Fig. 1.14 Blade failure under bending loading (Jinshui et al., 2013)

In of the Jensen et al. (2013), tested to failure a 1:1 scale, 34 m long, fiberglass-epoxy blade made of 1:1 scale. The turbine blade was subjected to bending in four sections using hydraulic cylinders acting in the extrados - intrados direction with the blade placed in a vertical position, as shown in Figure 1.15.

Fig. 1.15 Full-scale blade bending test in the extrados - intrados direction (Jensen et al., 2006)

Previously, a nonlinear finite element analysis of the tested blade was performed with *shell* and *brick* finite elements. From the analysis, it was predicted that the 13m long section of the blade (measured from the hub) would fail first, as shown in Figure 1.16.

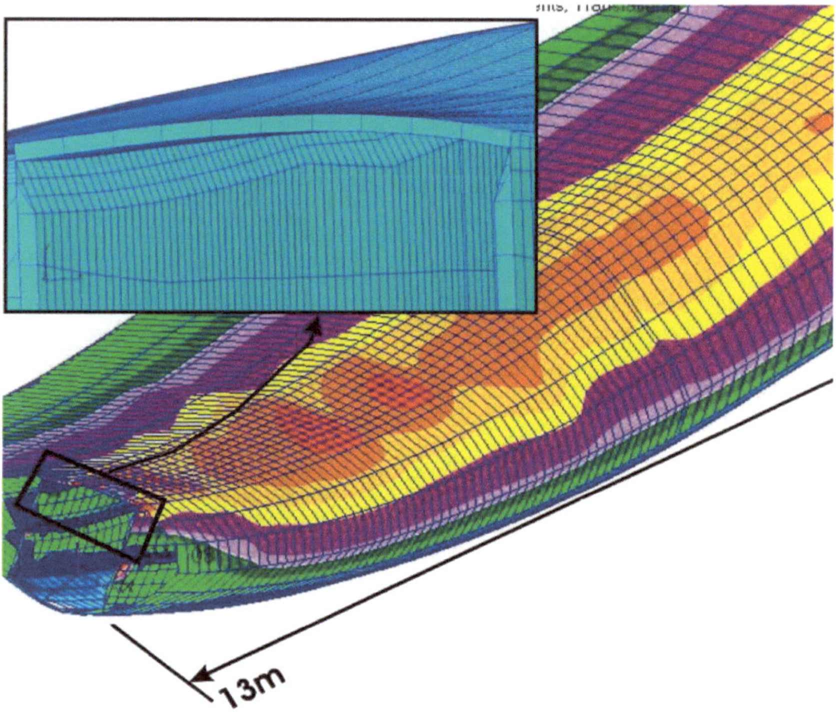

Fig. 1.16 Blade portion submodel 0-13m (Jensen et al., 2006)

The blades were tested statically and dynamically, simulating the stresses that would occur during 20 years of operation. During the tests, specific elongations (using strain gauges) and displacements on the blade surface were measured in seven sections of the blade to determine the spar ovalities, using digital comparators, as shown in Figure 1.17.

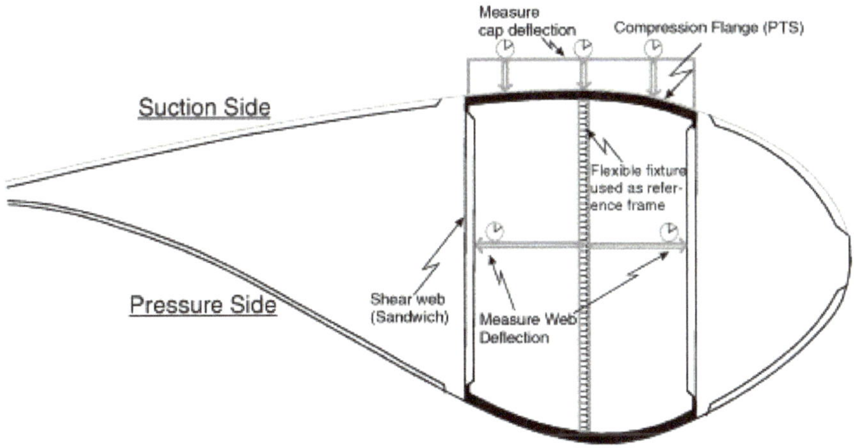

Fig. 1.17 Local displacement equipment in a blade section (Jensen et al., 2006)

FEM predicted that the failure mechanism would be initiated by the peeling of the skin, followed by its buckling leading to delamination. The tests carried out confirmed the results obtained by the AEF, as shown in Figure 1.18.

Fig. 1.18 Buckling collapse. (Jensen et al., 2006)

In of the Yadong et al. (2023), studied the influence of seawater infiltration on the composite material in the structure of a 1 MW wind turbine blade with a length of 8 m. It was shown that the strength of the composite material was substantially affected by seawater infiltration. An FEM analysis of the rotor using shell finite elements for the casing was done.

The AEF model showed a good agreement with the experimental determinations, the error being 4% between the measured and the calculated deflection. The elastic and mechanical characteristics of the composite material, affected by seawater infiltration, were then substituted into the AEF, estimating the negative impact of the environment on the turbine performance. By absorbing seawater, the elastic and mechanical characteristics of the composite material were affected. It was found that the stiffness, tensile strength and service life at variable loads decreased. However, the experimental results show that the blade that absorbed seawater was correctly designed and can take the maximum load provided.

To simulate the loading produced by the pressures distributed on the blade surface, the loads were applied using three servo-hydraulic actuators. The actuators applied force to the blade by means of large-area sleeves (one sleeve for actuator 1 and two each for actuators 2 and 3, the force being divided equally between the two sleeves by means of a cylindrical joint), as in Figure 1.19. Thus, 5 loading zones were realized, which allowed a better approximation of the blade load distribution during operation. The sleeves were made of high-density polyurethane foam and had the shape of the blade surface in the contact areas. The surface area of each sleeve was approximately 3000 cm2. Due to the large contact areas, no local crushing of the blades in the contact area with the sleeves was reported (Yadong et al., 2023).

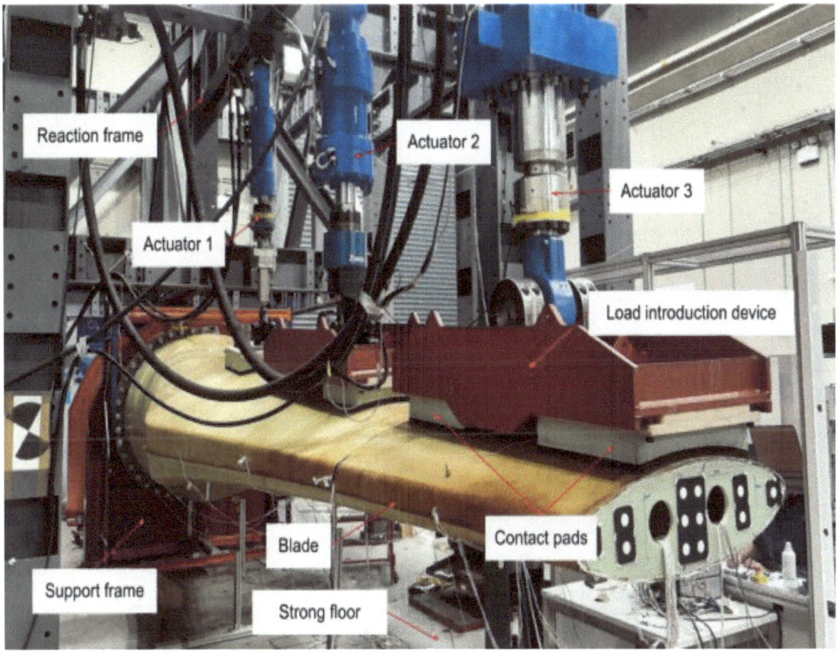

Fig. 1.19 Stand for bending blade testing (Yadong et at., 2023)

Few torsional tests of blades are known of Wang. et al. (2023) studied the torsion of a 5m long blade section taken from an in-service rotor. The torsion of the blade segment was performed using a stand designed for this purpose, as shown in Figure 1.20.

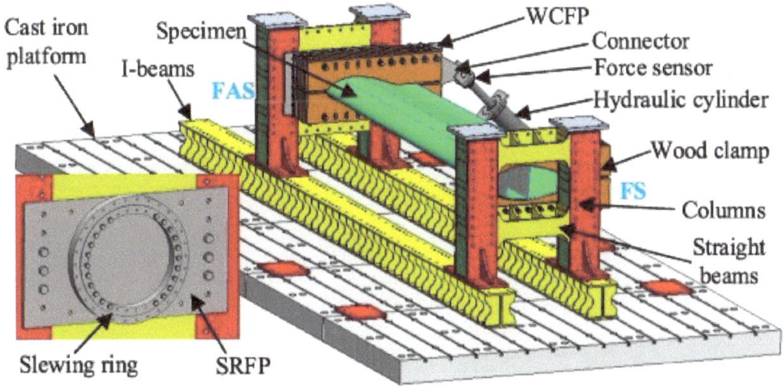

Fig. 1.20 Stand for torsion blade testing (Wang et al., 2023)

The segment ends of the blade were fixed using two wooden couplings, which have the blade profile on the inside and are assembled on the outside using a steel frame and bolted. A finite element analysis was performed with the program ABAQUS 16.0. Figure 1.21 shows the loading and discretization of the blade segment.

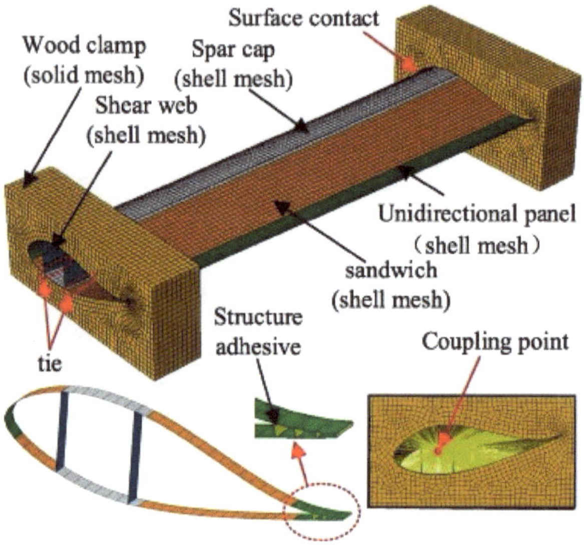

Fig. 1.21 FEM model of blade segment and simulation settings. (Wang et al., 2023)

The FEM results are shown in Figure 1.6. On the extrados and the intrados of the blade segment, tensile strain gauge marks were glued in the area of the adhesive joints. The loading was done by means of a hydraulic cylinder operated by a force sensor. The angle of rotation was measured using an angle sensor (clinometer) as in Figure 1.22.

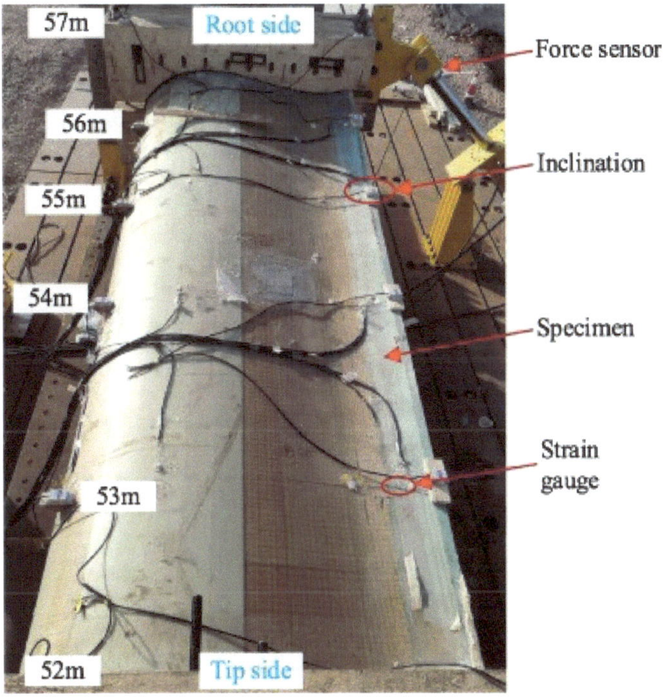

Fig. 1.22 Sensors set up in subcomponent test (Wang et al., 2023)

During the test, the modes of blade failure were investigated: debonding of adhesive joints, buckling in the trailing edge area, spar failure. It was found that the rotation of the sections is not directly proportional to the torsional moment. The rotations increase rapidly at the beginning and slower after the moment reaches 15 kN·m. Maximum tangential stresses are localized in the vicinity of the embedment. In the region of the stringer, negative motions were recorded, indicating compressive stress. Positive and negative stresses occur along the sandwich elements. Stress concentrations occur in the vicinity of wood couplings. Tests have shown that the highest tangential stresses occur in the stringers and sandwich elements. Tangential stresses are low in adhesive joints. The experimental results were in good agreement with those predicted by AEF. Larger errors occur after the initiation of some material failure.

Xiao (2017) investigated a 47 m long blade subjected to torsional buckling. The test was performed with the shovel placed in two positions, vertical and horizontal, respectively. The static tests were performed until the blade was damaged. The video recordings were analyzed frame by frame to observe the blade failure mode. It was found that the occurrence of local deformation led to the failure of the statically bent blade. The torsion affected the behavior of the blade, especially in the post-collapse stage. This could be explained by the fact that the torsional moment is much smaller than the bending moment. Figure 1.23 shows the loading mode of the blade, and Figure 1.24 shows the variation of bending moment and torsional moment along the blade length.

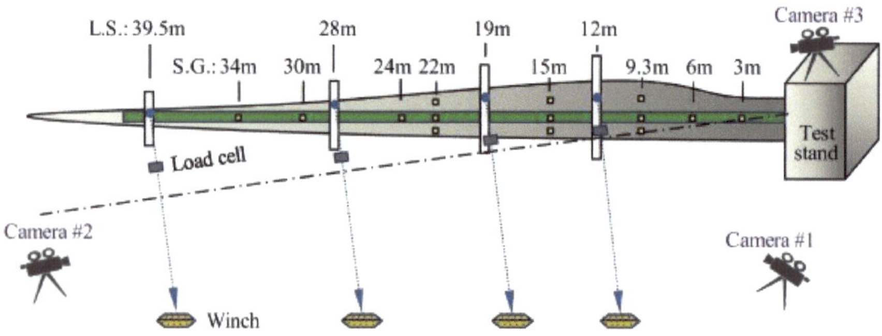

Fig. 1.23 Loading setup of the blade in the collapse test. (Xiao, 2017)

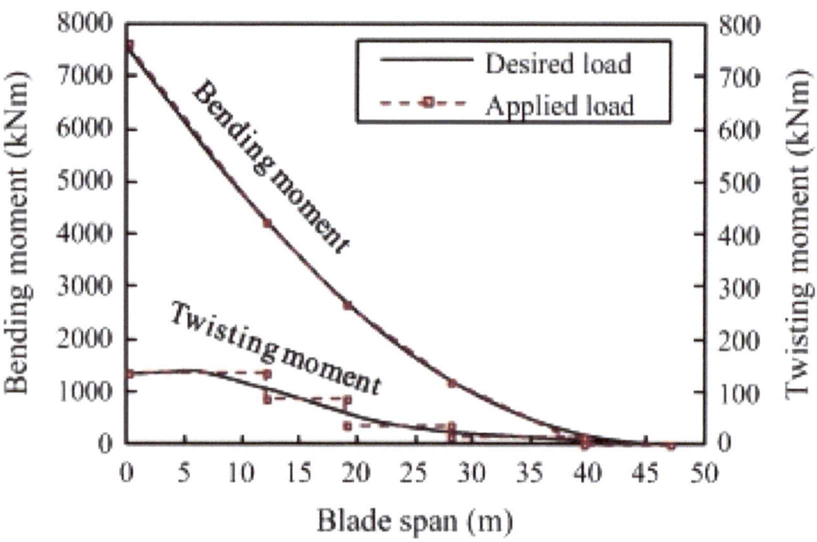

Fig. 1.24 Bending moment and twisting moment of the blade at 100% test loads. (Xiao, 2017)

The blade failure modes are shown in Figures 1.25 and 1.26.

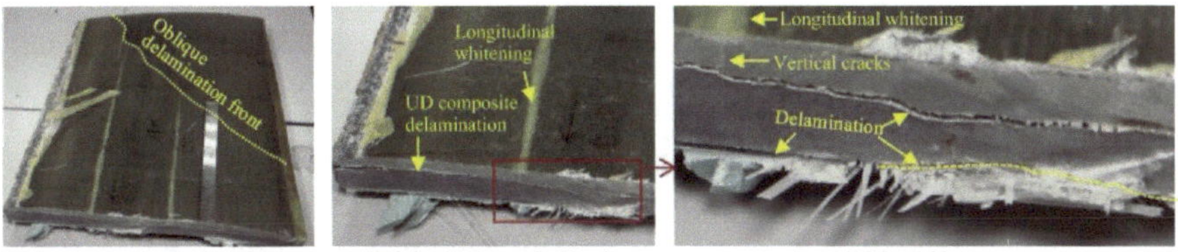

(a) Sample #2 to be examined (b) Delamination and whitening (c) Enlarged image

Fig. 1.25 Examination of the failure mode of the shell in the spar area (Xiao, 2017)

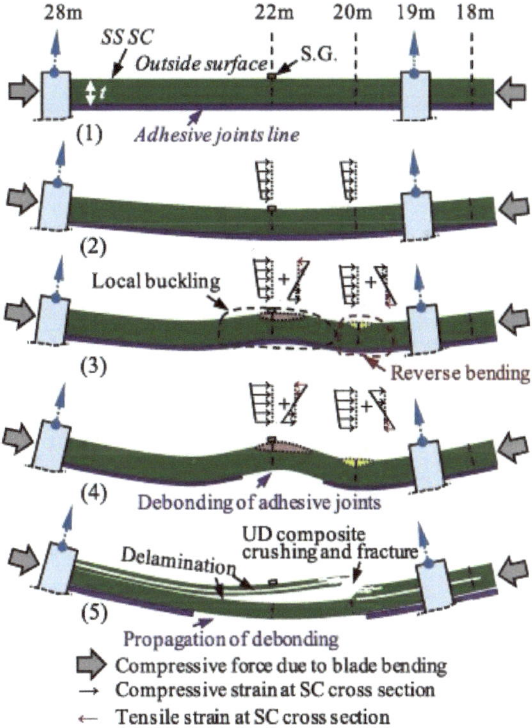

Fig. 1.26 Evolution of the shell failures in the area between the spars (Xiao, 2017)

The primary causes of blade failure were: longitudinal crushing due to compression followed by delamination of the skin in the area between the stringers on the extrados side. These two modes of failure resulted in the loss of the load-bearing capacity of the blade and determined its load capacity. Local buckling of the sheathing in the area between the stringers was the phenomenon that led to the final collapse of the blade. In the area of contact with the wooden couplings through which the blade is loaded, local crushing and stiffening occurs. The study did not reveal a major influence of torsion on the load-bearing capacity of the blade. However, torsion did contribute to blade failure in the post-collapse domain. The distribution of specific elongations was not uniform over the shell thickness in the area between the stringers, where failure is initiated due to longitudinal compression, followed by delamination. This shows the importance of analyzing the spatial stress state in the blade shell.

The failure of a full-scale blade under torsional bending, as shown in Figure 1.27, was presented in (Xiao, 2017; Zhang, et al., 2019).

Xiao, (2017) found that the failure of the GFRP blade in torsional bending was initiated by the occurrence of delaminations in the composite layer in contact with the spar as a result of compressive stresses, followed by local buckling. The same failure modes were predicted with FEA. To obtain the torsional bending composite test, the forces were applied with calculated eccentricity, as shown in Figure 1.28.

Fig. 1.27 Setup of a full-scale blade test under combined flapwise bending and torsion (Xiao, 2017)

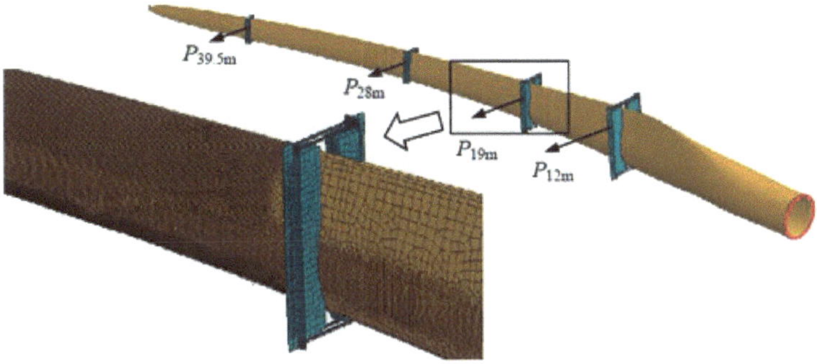

Fig. 1.28 AEF - Loading of the blade subjected to torsion with bending (Xiao, 2017)

During the lifetime of a blade (about 20-25 years), it is subjected to variable stresses, totaling about 10^9 cycles. The properties of composite materials degrade over time. It is important to understand the behavior of composite materials under variable stresses, but also of blades in full size. There are few studies on the fatigue failure of turbine blades.

In composite materials, fatigue failure occurs in three stages, depending on time (Marín et al., 2009):

1. In the initial period (stage 1), small damage to the matrix occurs, resulting in a small decrease in stiffness;
2. In the second stage, the damage increases linearly, over a long period of time, with a gradual decrease in stiffness. Delaminations and interlaminar detachments appear, and at the end of the stage, fiber breakage may also occur;
3. The third stage is characterized by the appearance of major delaminations and fiber breakage, which ultimately lead to material failure, as shown in Figure 1.29, (D'Amore et al., 2019).

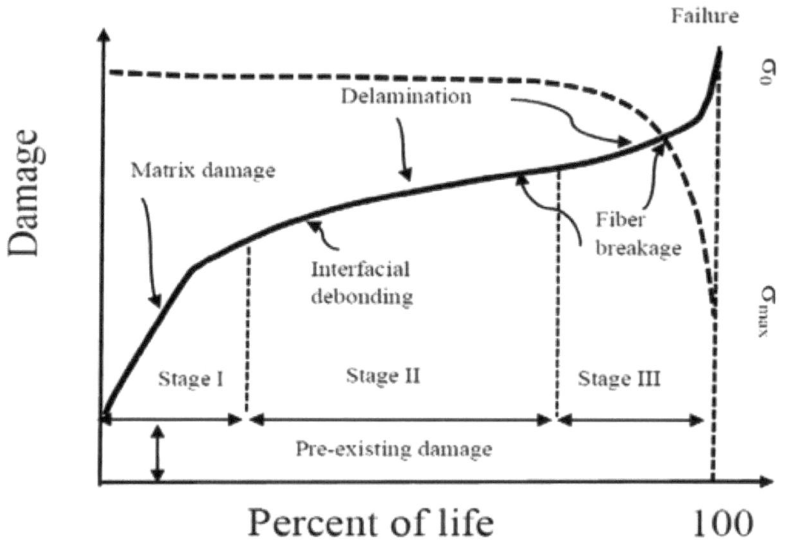

Fig. 1.29 The schematic of damage development and the corresponding strength evolution under constant amplitude loadings (D'Amore et al., 2019)

In most cases, micro-cracks initiated by blade fatigue first occur at the junction of the aerodynamic zone with the blade hub. In this area, there are stress concentrators: modification of the cross-section, holes for the bolts fixing the blade, joining of materials of different nature in the hub area (aluminum alloys with composite materials), existence of residual stresses in these materials. These micro-cracks propagate over time, as shown in Figure 1.30, and can lead to blade failure, as in Figure 1.31. The trailing edge area can be damaged by shear stress (Chen et al.,2020).

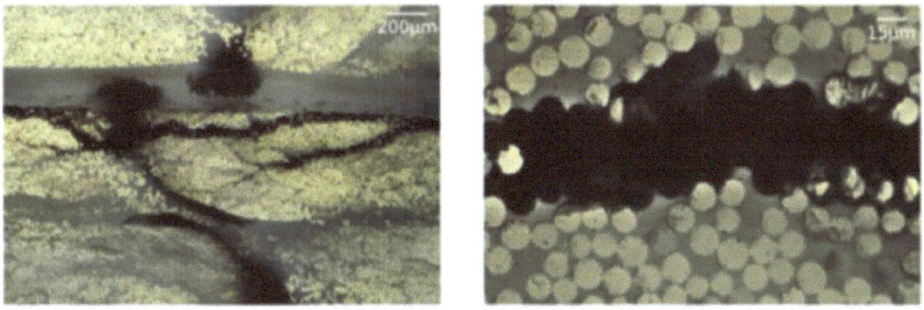

a) b)

Fig. 1.30 Microscopic observation of crack propagation through the matrix (a) and SEM detail (b) (Rocha, 2017)

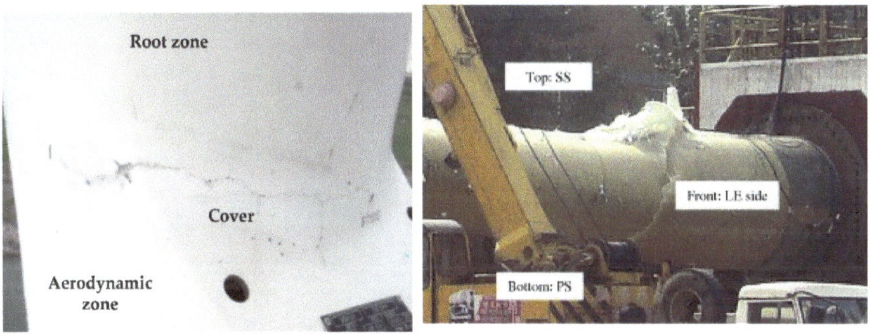

a) b)

Fig. 1.31 Cracking at the junction between the hub and the aerodynamic area (a), failure of the blade at the transition region. (b) (Marín et al.,2009)

Xiao (2019) studied the fatigue behavior up to 2 million cycles of a 47m long blade for a 2MW turbine. The tests were conducted according to the requirements of the international standard IEC 61400-23. Bending stiffness, natural frequencies and damping factor were checked periodically during the tests.

Initially, the blade was subjected to static loading, performed according to the above standard. The tests showed that the blade exhibited a linear response to the static loads and no blade damage was observed. The blade was embedded in the hub and an eccentric rotor driven by an electric motor was mounted on the blade at a distance of 39.5m. This excited the blade to vibrate freely. Accelerometers were glued on the shovel in the sections at 19m, 29m and 39.5m from the embedment as in Figure 1.32.

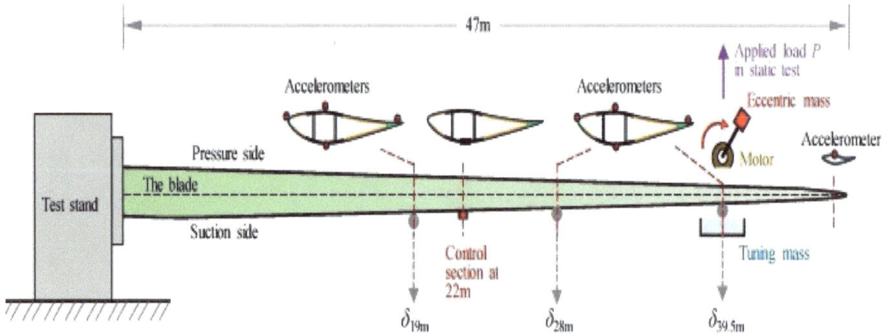

Fig. 1.32 Test setup for fatigue testing of blades (Xiao, 2019)

FEA and experimental results are in good agreement, as in Figure 1.33.

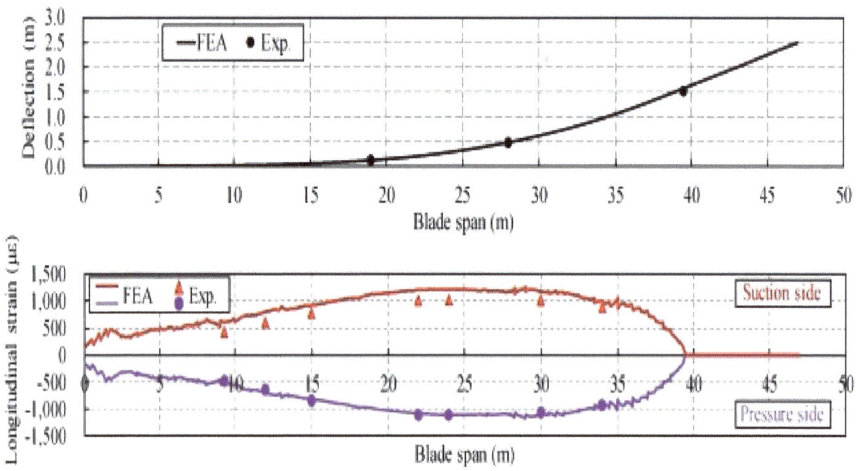

Fig. 1.33 The deflection and longitudinal strain in the spar cap of the blade loaded statically at 39.5 m (Xiao, 2019)

During the fatigue test, it was observed that the blade stiffness does not change significantly over the length 0-19 m, but over the distances 0-28 m and 0-39.5 m it decreases considerably, respectively, as shown in Figure 1.34.

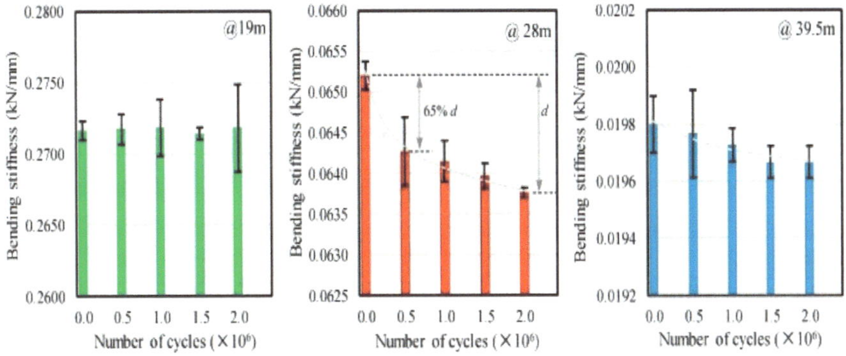

Fig. 1.34 Degradation of bending stiffness of the blade during fatigue test. (Xiao, 2019)

The change in the blade natural frequencies after the fatigue test was insignificant (< 1 %). In contrast, the damping factor showed increases of up to 20 %, both for bending stress in the intrados - extrados and leading edge - trailing edge directions, respectively, and for torsional stress.

After fatigue testing, the blade was examined for defects. Few flaws were found, which included sheathing bleaching and cracking in the core of the sandwich composite, as well as sheathing tears in the transition region as in Figure 1.35. These defects are local and are considered to have negligible effect on the structural strength of the blade.

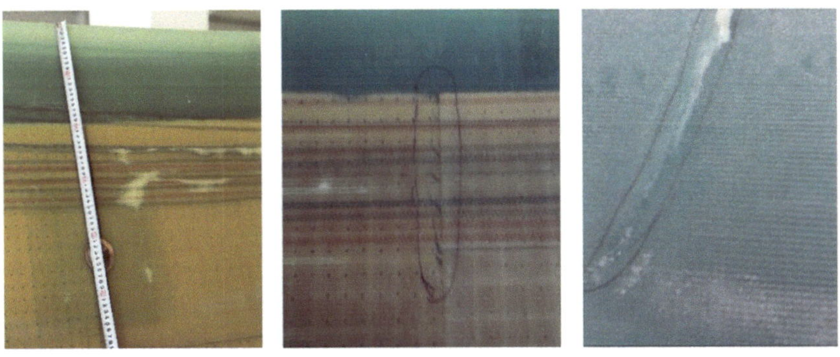

Fig. 1.35 The observed blade damages after the fatigue test (Xiao, 2019)

In of the Liu et al. (2022), a FEA of 59.5 m long blades from 2MW wind power plants for different wind speeds was presented. The stress state (maximum normal stresses were 30-50 MPa in the hub area), strain and displacements in the blade were outlined, and the fatigue behavior of the blades was analyzed. By introducing a composite degradation model, the evolution of the composite characteristics as a function of the number of cycles was predicted.

The blade endurance was evaluated by combining three methods: the Rainflow counting algorithm, Wohler diagrams and the cumulative degradation principle. It was estimated that the life of the blades under the studied operating conditions would be 19-22 years. Figure 1.36 shows the Wohler diagram for the blade made of different composite materials.

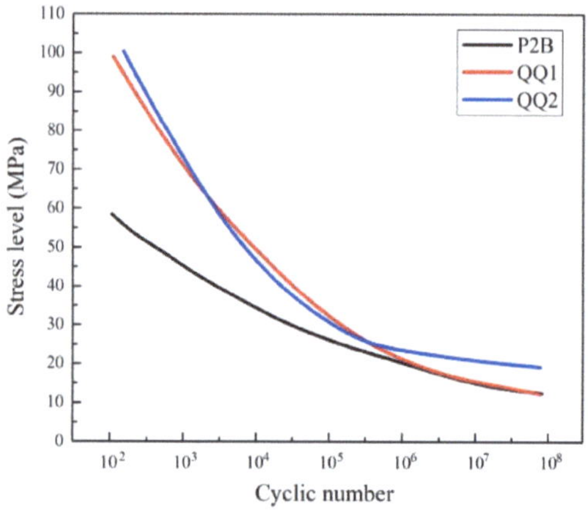

Fig. 1.36 The S–N curves for composite materials P2B, (C-glass hybrid laminate), QQ1(GFRP), QQ2 (GFRP), (Liu et al., 2022)

In of the Fremmelev et al. (2022), the fatigue testing of a 52 m long blade in which different types of defects were initially introduced to follow their evolution as a function of the number of cycles is presented. The structural integrity of the blade was monitored with different sensors: strain gauges, accelerometers, acoustic emissions, digital image correlation (DIC) and piezoelectric sensors for vibration detection. The study highlighted the need to use different types of sensors for monitoring the structural integrity of blades. Out of the defects initially introduced, the appearance and fatigue propagation of a crack were found only in the area of a cut in the stringer on the intrados side. A delamination developed at the crack tip, but this propagated more slowly to the portion of the stringer on the top surface. The shovel-mounted sensor system was able to detect the crack propagation and this was considered successful.

1.4 Blade monitoring during operation

In recent years, wind turbines have developed significantly. Due to the increasingly long blade designs, much more complex monitoring and control are required.

To protect wind turbines from damage, proper monitoring is necessary, which increases the efficiency of the installation. Because of their size, complex loads, and the harsh environmental conditions in which turbine blades operate, it is impossible to test them under real conditions. For this reason, and also to prevent accidental failures, monitoring the blades during operation plays an extremely important role.

Non-destructive testing (NDT) techniques are increasingly used to monitor the structural integrity of blades during operation. In addition to providing useful data for design, these techniques can identify the causes of defect occurrence and propagation and can help prevent catastrophic blade failures. Below are the main NDT techniques used for monitoring blades during operation. Moreover, structural integrity monitoring of the blades (SIMB) leads to a reduction in maintenance work and extends their service life. Gholizadeh (2016) classified the NDT techniques for MISP into contact and non-contact techniques. At present, the main NDT techniques used for MISP are based on strain measurement, acoustic emission, ultrasound, vibrations, thermal imaging, DIC, etc.

Strain measurement methods offer the advantage of being minimally intrusive and capable of continuously monitoring the Structural Integrity Monitoring of Blades (SIMB) over long periods of time. Their disadvantage is that the accuracy and sensitivity of the measurements depend on the distance between the sensors and the propagating defects (Ozbek, 2013). These techniques include strain gauges bonded to the blade surface and fiber Bragg grating (FBG) diffraction sensors embedded into the composite material during manufacturing. Strain gauges have the drawback of collecting information from a small surface area; consequently, a large number of sensors are required for SIMB. Since they are surface-mounted, they must be well protected from environmental effects. Both conventional strain gauges (with a single grid) and strain gauge rosettes can be used. The advantages of these sensors include low cost, very small size, and high accuracy and sensitivity.

There are strain gauges such as CPA01-02 and CPC03 (Measurement Group – Vishay) used to monitor crack propagation speed. They can be mounted near the tip of a crack, which, as it propagates, cuts through several parallel-mounted filaments, resulting in an increase in the sensor's electrical resistance. Although their usefulness in fracture mechanics is evident, no known applications of these sensors for SIMB have been reported.

FBG sensors consist of a very thin optical fiber that contains one or more regions (approximately 10 mm in length) with Bragg diffraction gratings. They are essentially non-intrusive and can be embedded into the composite material during its manufacturing.

A single sensor can extend along the entire length of the blade and may include up to 40–50 diffraction gratings, which are used to measure strain. FBG sensors can operate up to a maximum strain of ε_{max}= 0.5%, while conventional constantan strain gauges with a grid can reach ε_{max} = 2%–20% (depending on the model, grid size, and adhesive used).

Temperature variations introduce significantly larger errors in FBG sensors compared to electrical resistance strain gauges (ERSGs), and these errors must be corrected. However, FBG sensors are immune to interference from varying electromagnetic fields.

HBM has fatigue tested FBG sensors with specific strain up to ±1000μm/m and found no change (sensitivity and wavelength unchanged) after 10^7 cycles. Under similar conditions, strain gauge showed a zero drift of 10μm/m at the end of the tests. FBG sensors can withstand up to significantly larger specific strain than TER (±2000μm/m).

Wu et al. (2017) proposed a new type of FBG strain gauge for MISP (Wu, et al., 2015). Lee et al. proposed a new method for optimizing the placement of strain gauge on the blade surface.

Researchers such as Bang et al., (2011), Guo et al., (2007) and Schroeder et al., (2005) have used FBG technique for monitoring deformations in dynamic regime in both land and offshore mounted turbine blades and found good accuracy, stability and durability of this type of sensor.

The acoustic emission method uses microphones glued to the surface of the blade, which can detect the sounds produced when defects occur (matrix cracks, fiber breaks, crack propagation, plastic deformation, etc.). The method can determine the location of crack occurrence and allows crack propagation monitoring.

Tang et al., (2016) and Zhou et al., (2014) used acoustic emission for MISP and have shown the capabilities of the method to detect the occurrence, localization and propagation of cracks in blades. However, the method can be affected by signal interference. Removing the interference significantly increases the cost of the measurements.

The ultrasonic testing method uses an ultrasonic wave that first propagates through the material and then the sensor detects the reflected wave. Depending on the material, different modes can be distinguished: reflection, attenuation and transmission (Amenabar et al.,2011). Typical defects that can be detected by the US method are: delaminations, defects in the adhesive layer, resin poor areas, etc. The method is fast, reliable and it is widely used for inspection of composite materials (Yang, 2013).

Marquez et.al. (2020) shows how to install ultrasonic sensors on the shovel and on the tower (reflection method), as shown in Figure 1.37.

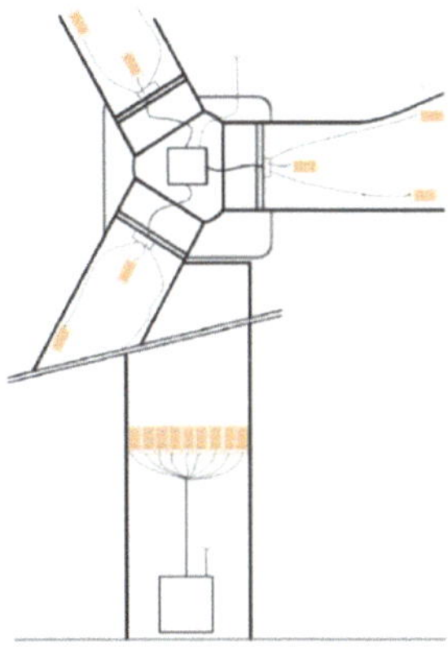

Fig. 1.37 Location of ultrasonic sensors on blades and tower (Marquez et al., 2020)

Studies on ultrasonic testing for MISP have been conducted by Marquez et al, Tiwari et al who detected internal cracks and damage to the blade surface, which ultimately caused material debonding, (Marquez et al, 2020; Tiwari et al,

2018). Other researchers such as (Brett et al., 2018; Habibi, et al. 2015) have studied defects such as internal delaminations and also measured the adhesive thickness by Phased Array method. In addition, this method can also be applied to detect the defects of offshore wind turbine blades. However, ultrasonic testing requires a long time for signal collection, and the data processing is also complex, which leads to the delay in fault assessment. Therefore, future research on artificial intelligence algorithms can improve data processing capability (Márquez et al., 2012).

The thermoelastic method is based on the measurement of the temperature variations that occur at the surface of a body as a result of a change in the stress state. A change in the stress state of a volumetric element made of isotropic material also produces a change in its temperature proportional to the first invariant I_1, which is equal to the sum of the principal stresses. The infrared (IR) cameras used must have a sensitivity of the order of 0.01-0.1°K. Using them and dedicated programs, it is possible to construct a "map" of the variation of the temperatures and the sum of the principal stresses on the surface of the structure. The method uses thermal imaging (IR) and it is mainly applied to detect the change of thermodynamic properties of wind turbine blades by scanning the surface. When micro-cracking of the material occurs, it causes a local temperature variation, which can be used to detect and evaluate the failure.

Doroshtnasir et al. (2016) used remote thermal imaging testing and calculated the value of blade temperature differences. It was found that the temperature is high near the hub and, as a result, there is the highest possibility of failure in that area. Avdelidis et al. (2013) applied the same NDT method but in infrared and were able to detect the internal defects. However, the temperature change caused by damage is delayed and it is easy for the ambient temperature to cause interference in the detection process.

Currently, a large number of innovative detection and monitoring systems based on the digital image correlation (DIC) method have been developed and applied in the MISP field. DIC has the following advantages: it is a zonal (full field), non-contact method; it is not necessary to know in advance the areas of maximum strain; it can be applied to flat (single chamber version) or curved (dual chamber) surfaces; it can determine the areas of maximum specific strain and indicate their numerical value; it can provide data in static or dynamic mode; it is easy to use; it can be applied to a wide range of materials. The disadvantages of the method are: it allows the determination of displacements and deformations only at the surface of bodies; it uses expensive equipment and software; the accuracy of the determinations depends on the quality of the images.

This method uses a series of images taken with high-resolution cameras during rotor operation to monitor deformations occurring on the blade surface (Yang, 2012). Other variations of the method make use of drones (Wang, 2019) or mobile robots that move on the blade surface and that are equipped with cameras or video cameras (Franko, 2020), as shown in Figure 1.38.

A new video technology for monitoring blade deformations during operation has been proposed by Yang et al. This technology requires two cameras to obtain stereoscopic images and the gluing of grids on the blade surface, as shown in Figure 1.38 a. This measurement method has been shown to have good accuracy and reliability.

Poozesh et al. (2014) have used a new optical method for the early detection of defects occurring in turbine blades. The method uses a pair of cameras to take stereoscopic images with the help of which specific deformations on the blade surface are determined. The method shows good accuracy, but the calculation process is rather complex and needs to be improved.

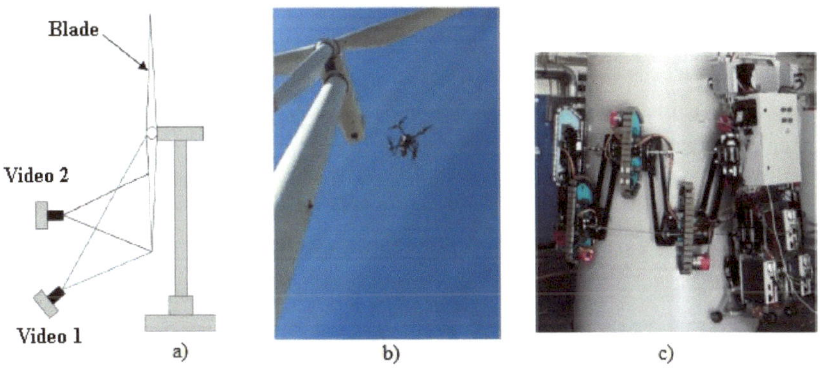

Fig. 1.38 a) Machine vision system (Yang et al. 2012); b) Data collection using drones (Wang et al., 019); c) Examination using mobile robots (Franko et al.,2020)

Although the technologies presented above are widely used, each of them has certain shortcomings, as shown in Table 1.5. A combination of several NDT methods is required to obtain accurate results.

Tabel 1.1 Advantages and disadvantages of NDT methods used for MISP (Hill, 1997; Kim, 2013; Qiao, 2015; Amezquita, 2016; Sanati, 2004)

Test method	Advantages	Disadvantages
Acoustic emission	Can locate cracks or delaminations caused by fatigue, as well as their development	Can be affected by environmental noise, data processing is complex and the cost is high
Ultrasound	It can assess the size and location of defects, etc.	Long-term signal acquisition and processing is complex and delays data processing.
Deformation measurement	Allows continuous monitoring of turbine blades	The sensitivity depends on the distance between the sensor and the monitored defect.
Thermography (thermal imaging)	Intuitive and efficient method	The ambient temperature can affect the accuracy of the measurements.
Optical methods (machine vision))	Low cost, safe, real-time data delivery to wind farm personnel.	Accuracy depends on the quality of the images and cannot explain the degradation mechanism of the material.

Mads et al. (2022) performed fatigue testing on a 52-meter blade, in which a series of defects had been introduced, whose evolution was monitored using several NDT methods, as shown in Figure 1.39. Following the tests, several conclusions were reached:

- Using strain gauges, the presence and propagation of defects in their vicinity could be detected. The recorded strains did not vary significantly during crack propagation. This method can be useful for SIMB in critical areas;

- The simultaneous occurrence of multiple defects makes result interpretation through acoustic emission more difficult. By using multiple sensors placed near the defect area, it can be located and its propagation can be monitored. The area from which the sensors can capture the signal is relatively limited due to signal attenuation during propagation through the composite material. Tests have shown that the efficiency of the sensors is limited to a distance of approximately 0.5 meters;

- The active blade vibration monitoring system was realized using 3D accelerometers. Using the data recorded with the accelerometers, it was possible to monitor crack propagation. This technique proved to be more sensitive to smaller flaws. It was possible to detect the size, localization and propagation of flaws.

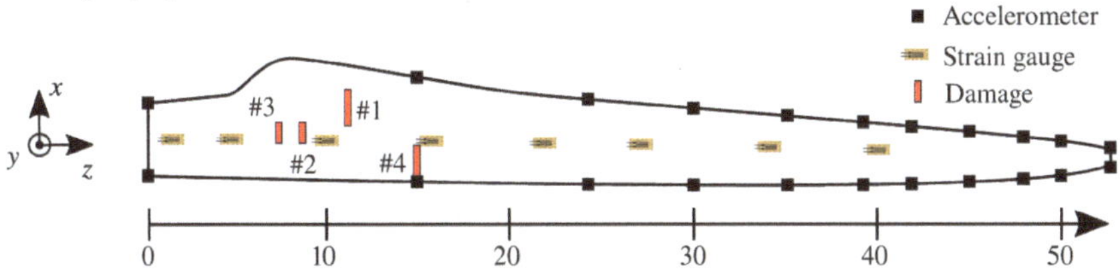

Fig. 1.39 Spanwise and edgewise placements of investigated damage cases, including global unidirectional SG and triaxial accelerometer placement. (Mads et al.,2022)

Gomez et al. (2016) used acoustic emission and strain gauges for the MISP of models and full-scale blades, respectively, as shown in Figure 1.40.

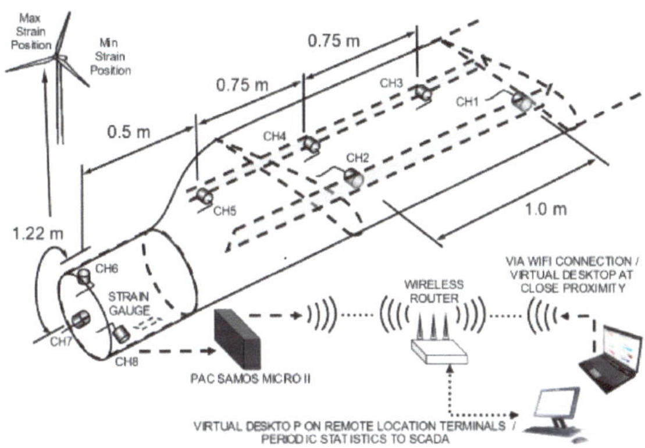

Fig. 1.40 Placement of microphones for acoustic emission and strain gauge on the blade (Gomez et al. 2016)

1.5. General conclusions to Chapter 1

1. During operation, the wind turbine blade is subjected to compound stresses with distributed forces: bending in two planes, axial stresses and torsion. These complex loads cannot be faithfully reproduced during blade testing. However, it is important that the blade tests reproduce as closely as possible the loads occurring during blade operation;
2. Due to the complexity of the shape, structure and loads, blade design can only be approached with numerical methods (FEA). Even in this case, FEA encounters some difficulties due to the diversity and complexity of the materials used in the blades and their dimensions. Due to power and computational time limitations, the blade will be discretized into a limited number of finite elements of larger dimensions, but which are still small relative to the blade dimensions;
3. In order to determine the strength limit of the blades, they are tested in static tests, usually in bending in the extrados - intrados direction and in the leading edge - trailing edge direction, respectively. Because of the complexity of the loadings, few tests of the blades have been carried out at compound stresses (only torsional bending), although their importance is obvious. Testing can be done on models (using similarity criteria) as well as on real blades;
4. During operation, the blade is subjected to variable stresses. This necessitates fatigue testing of both the blade materials and the entire blade (scale 1:1). Fatigue tests cannot be performed on blade models. Due to the size of the blades, duration and complexity of the loads, only a small number of blades can be fatigue tested, although tests at varying stresses require a large number of specimens/structures tested;
5. Monitoring the structural integrity of the blades during operation is carried out both to prevent blade breakage and to obtain design data. This monitoring can be carried out by various methods: strain gauge, FBG, acoustic emission, ultrasonic, DIC, thermoelastic method, etc. More accurate results can be obtained by combining the results obtained by several methods.

CHAPTER 2. FINITE ELEMENT ANALYSIS OF SMALL-SCALE WIND TURBINE BLADES

2.1 The scope of finite element analysis

Due to the complexity of the profiles and loads applied on the wind turbine blades, analytical approaches can only provide approximate results. For this reason, rotor blades are designed using numerical simulation methods, in particular methods using finite element analysis (FEA).

The finite element method (FEM), therefore, becomes an essential tool for the study of loading conditions, deformations, displacements and stresses of wind turbine blades. An important advantage of FEA is that this method can simulate both the actual loads during operation as well as the concentrated force and moment loads that occur when testing rotor blades. For the finite element analysis in this book, the composite material was considered homogeneous and orthotropic. This assumption is accepted by many researchers and leads to sufficiently accurate results.

The composite material used exhibits orthotropy in the plane, i. e. on the surfaces to which the strain gauges have been bonded.

This property reflects the differences in stiffness and mechanical behavior in different in-plane directions, caused by the orientation of the fibers within the composite structure. Therefore, incorporating these characteristics into the finite element model allows for a more realistic prediction of the structural response and a better correlation with the experimental data obtained during testing.

2.2 Wind turbine blade design and construction

The turbine blade was designed and manufactured from an E-glass fiber composite, incorporating a reinforcing system to ensure both strength and durability under operational loads. E-glass composites are widely used in wind turbine blade manufacturing due to their favorable combination of high tensile strength, corrosion resistance, and relatively low density, which together contribute to an excellent strength-to-weight ratio. The reinforcement system within the composite layup was specifically arranged to optimize load transfer and enhance resistance to both bending and torsional stresses, which are critical for blades subjected to variable aerodynamic and centrifugal forces.

The total blade length is 1721 mm, maximum chord 273.71 mm and chord tip 81.54 mm, as shown in Figure 2.1.

The blade is fixed to the rotor by means of a 100 mm diameter coupling, Figure 4.2. A NACA airfoil profile was used for structural performance. In order to achieve increased structural strength and stiffness at 0.286R (R- total blade radius = distance from rotor center to blade tip), the same NACA airfoil profile was applied for both the upper and lower surfaces, as shown in Figure 2.2.

This design choice, as reported by Savin A. et al. (2018), enhances the blade's resistance to bending moments generated in this critical mid-span region, where aerodynamic loads and centrifugal effects are both significant.

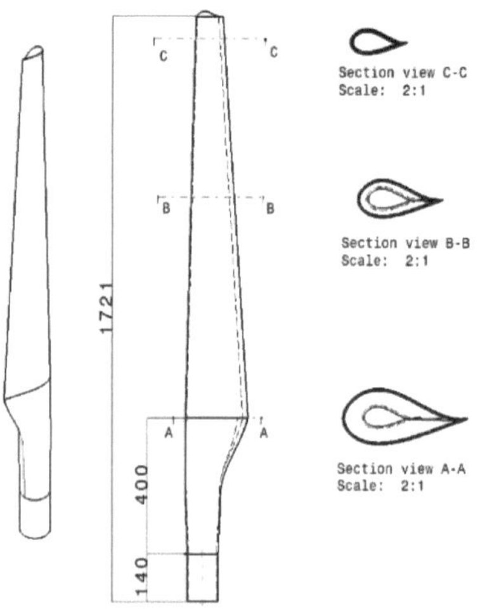

Fig. 2.1 Design of wind turbine blade model

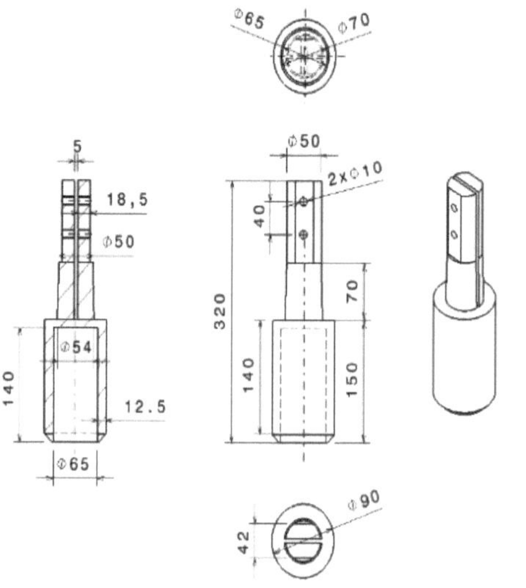

Fig. 2.2 Design of the blade hub

2.3 Pre-processing the wind turbine blade model

The configuration of the GE1.5sle wind turbine blade was determined by studying the parameters using FEM. The finite element analysis was performed in ANSYS Academic R17.2 software in the Static Structural workbench.

The discretization network of the three components of the blade assembly (the blade itself, the stiffening element, the hub attachment element) was obtained by combining a general parameterization with a particular/individual parameterization on each component.

In the general parameterization, the curvature-type dimension function was specified (in areas where the curvature is more pronounced, the discretization network will be finer), with a normal curvature angle of 24 degrees, slow transition, an average smoothing of the network and a cell growth rate of 1.1. In the parameterization performed on each component of the assembly, the maximum size of the discretization network element/cell was specified as follows:

1. hub clamp: elements with maximum dimensions of 4 mm;
2. stiffening element: elements with maximum dimensions of 4 mm;
3. the blade itself:

    - elements with maximum dimensions of 5 mm, growth rate of 1.05;
    - the trailing edge on the wingspan on the active side of the blade with 1.5 mm elements and a growth rate of 1.03;
    - the edge in the transition area between the inactive and active parts of the blade in the trailing edge area with 2 mm elements and a growth rate of 1.05.

During the simulations, two materials were used for the assembly components:

1. GFRP - for the blade itself and the stiffening element (**the material characteristics determined in Chapter 2, from the book Testing of composite materials used in wind turbine blades, were introduced into the GFRP material properties in Ansys**);
2. Steel (default from ANSYS 17.2) - for the hub fastener.

The discretization mesh, as shown in Figure 2.3, of the blade assembly was made exclusively of tetrahedral elements with a number of 874815 elements and 1466508 nodes.

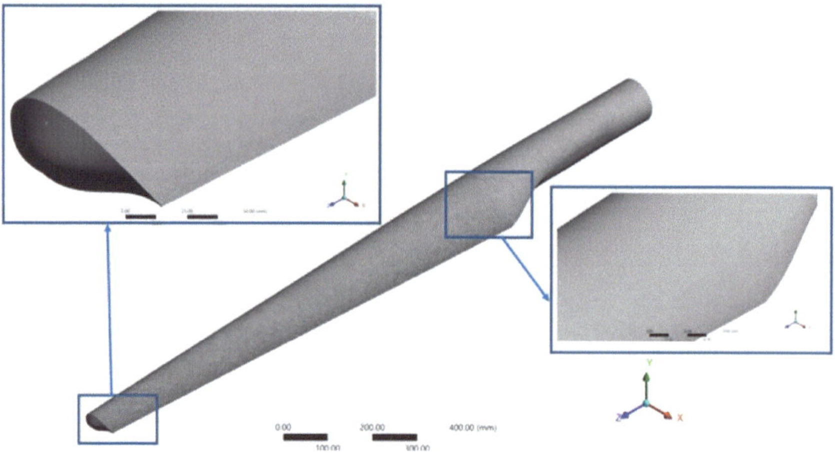

Fig. 2.3 Finite element mesh of the blade

The post-processing of the FEM results was performed for 5 different loading cases. As a result of the experiments carried out within the PCCA 2013-4-0656 project, contract no. 59/01.07.2014, it was estimated that the maximum load to which the turbine blade model can be subjected is: F=1000 N (applied to the blade tip) and M_t= 170660 N·mm (applied to the blade tip).

1. Bending F= 1000 N;
2. Clockwise torsion M_t= − 170660.646 N·mm;
3. Anticlockwise torsion M_t= 170660.646 Nmm;
4. Bending + Clockwise torsion F= 1000N + M_t= −170660.646 N·mm;
5. Bending + Anticlockwise torsion F= 1000N + M_t= 170660.646 N·mm.

2.4 Case 1: Finite element analysis for bending: F= 1000 N

The loaded state is achieved by clamping the hub and applying a force of 1000 N in the intrados-extrados direction on the blade tip surface, as shown in Figure 2.4.

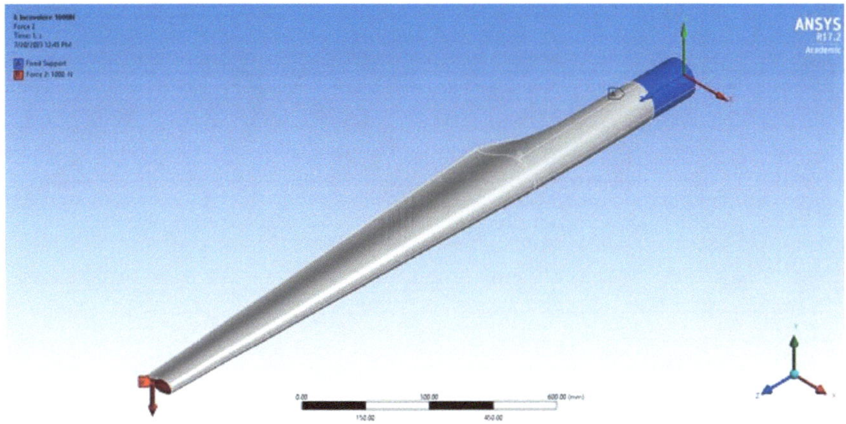

Fig. 2.4 Bending load condition

The distributions of the von Mises stress

The results of the finite element analysis are presented with respect to a coordinate system attached to the blade, with the origin at the base of the blade, the center on the hub symmetry axis, the Z-axis along the blade, the X-axis in the trailing edge - leading edge (positive) direction and the Y-axis in the extrados - intrados (positive) direction, for all five simulation cases.

Thus, the force applied in the extrados direction is 1000 N. For finite element analysis, the coordinates are given in terms of the same coordinate system.

Figure 2.5 shows the von Mises equivalent stress distribution, the maximum being 60.318 MPa, which occurs on the inner surface of the extrados. The coordinates of the point at which the maximum von Mises stress was obtained according to the attached coordinate system are X: -6.282 mm, Y: -14.13 mm, Z:1425.93 mm.

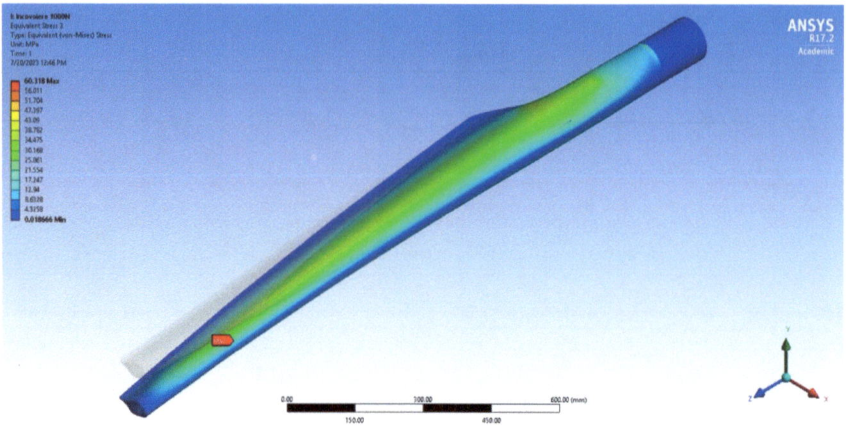

a)

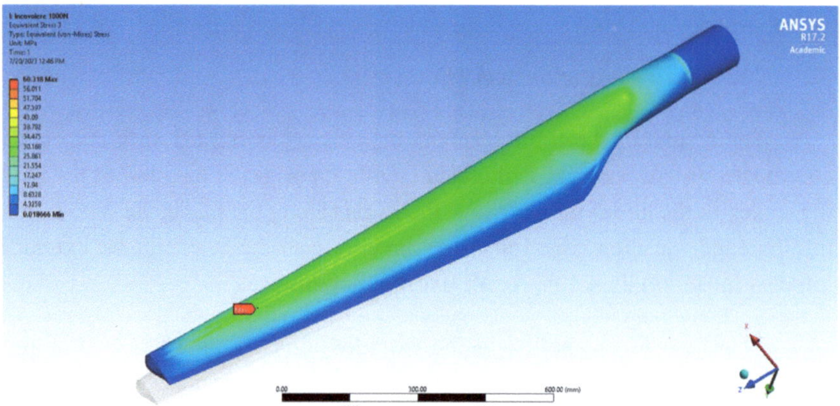

b)

Fig. 2.5 The distributions of the von Mises stress: on the intrados a) and, respectively, on the extrados b) of the blade

Figure 2.6 shows the distribution of the von Mises equivalent stresses on the inside of the blade, and Figure 2.7 shows, in cross-section, the distribution of the von Mises equivalent stresses in the plane, where the maximum von Mises equivalent stress was obtained (the maximum equivalent stress is highlighted with a red marker).

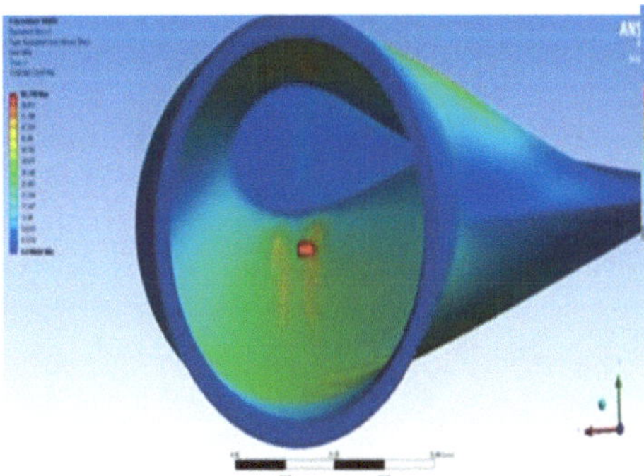

Fig. 2.6 View from the hub

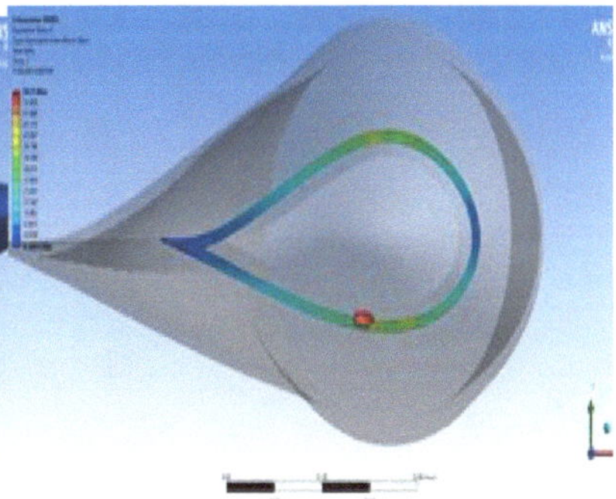

Fig. 2.7 Plan view of the von Mises stress distribution where the maximum stress was obtained

Figure 2.8 shows the distribution of the maximum von Mises equivalent stresses at 130 mm from the blade tip, $\sigma_{ech\ max}$=35.16 MPa. Figure 2.9 shows the distribution of the maximum equivalent stresses at 230 mm from the blade hub, $\sigma_{ech\ max}$=13.84 MPa.

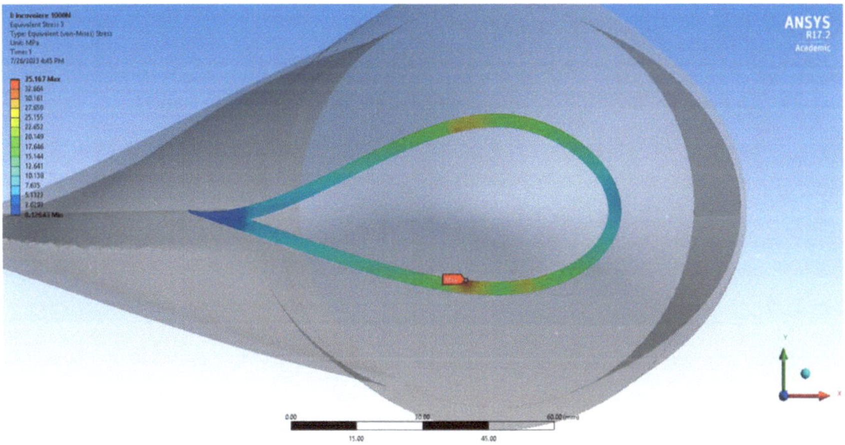

Fig. 2.8 Maximum equivalent stress distribution at 130 mm from the blade tip

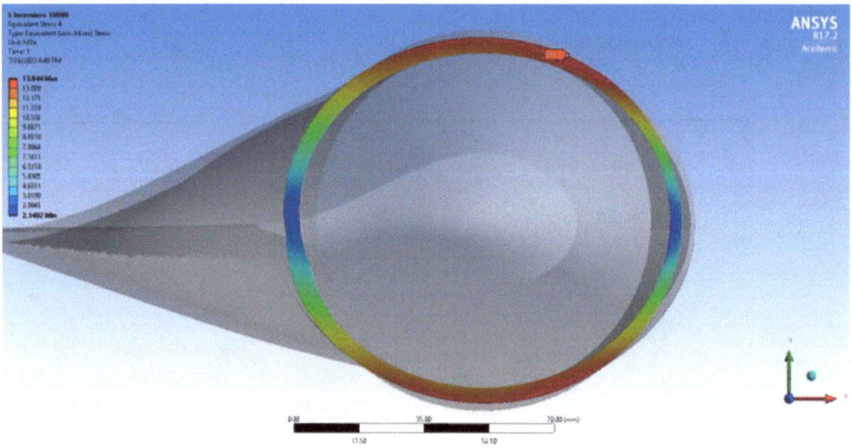

Fig. 2.9 Maximum equivalent stress distribution at 230 mm from the blade hub

Distribution of maximum principal stress

Figure 2.10 presents the maximum principal stress map, which is 41.83 MPa and occurs on the inside surface of the blade at the coordinates X: -5.729 mm, Y: 23.85 mm, Z: 1254.23 mm, also located on the inside surface of the blade skin.

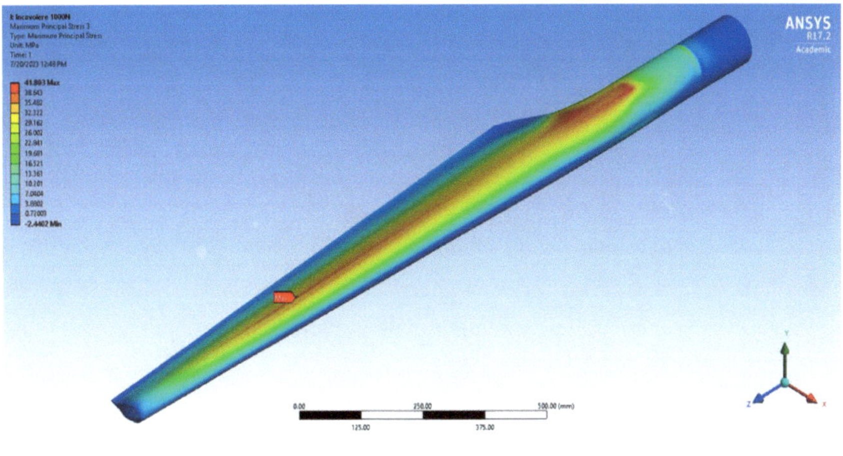

a)

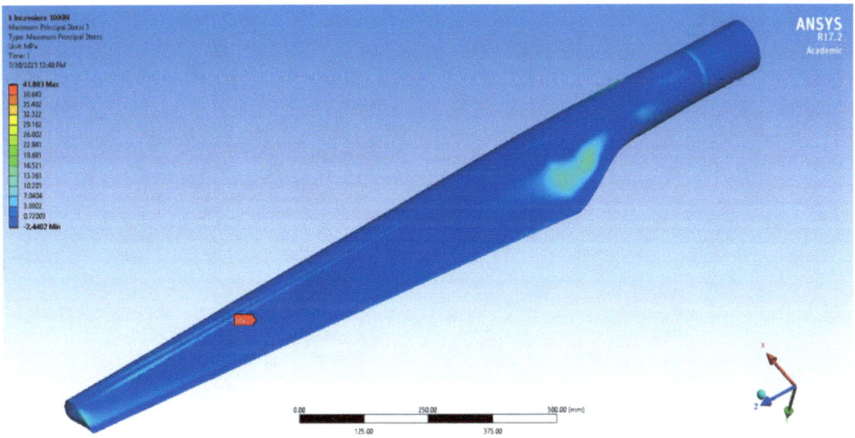

Fig. 2.10 The maximum principal stress: on the intrados a) and, respectively, on the extrados b) of the blade

Figure 2.11 shows the maximum principal stress - plan view. Figure 2.12 shows the maximum principal stress at the point where the maximum von Mises stress $\sigma_{1\ max} = 38.03$ MPa occurred.

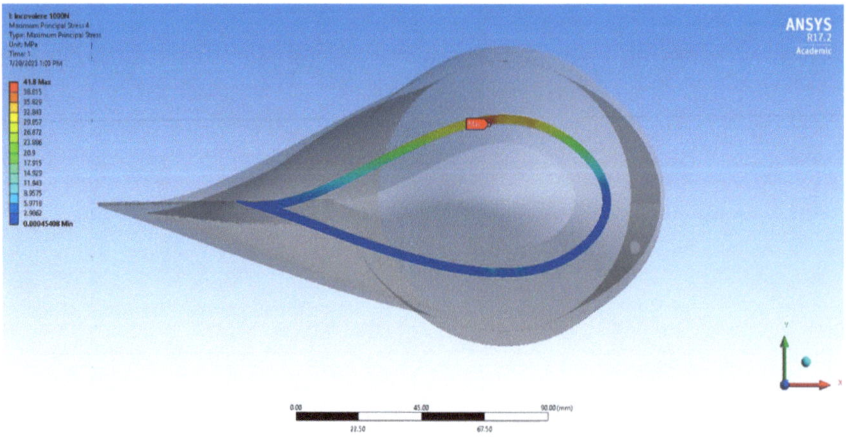

Fig. 2.11 Plan view with the von Mises stress distribution where the maximum stress was obtained

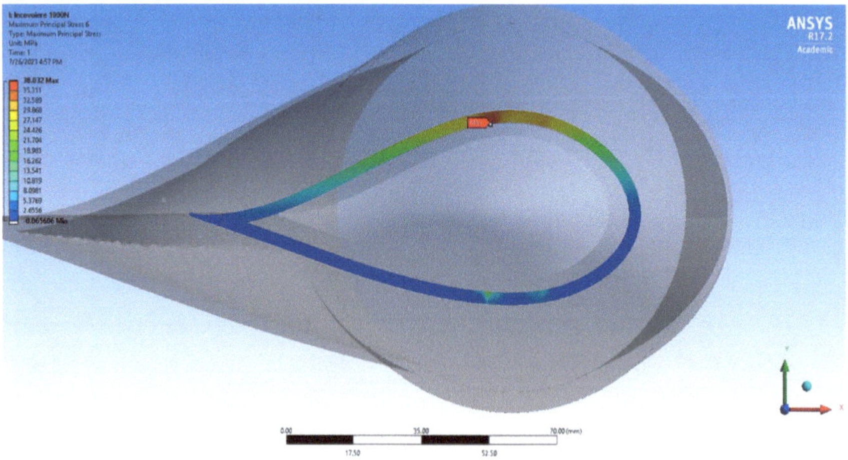

Fig. 2.12 Maximum principal stress at the point where the maximum von Mises stress occurred (X:-6.282 mm, Y:-14.13 mm, Z:1425.93 mm)

Figure 2.13 shows the maximum principal stress at 130 mm from the blade tip σ_{1max} = 25.92 MPa. Figure 2.14 shows the maximum principal stress at 230 mm from the blade hub $\sigma_{1\ max}$ = 13.82 MPa.

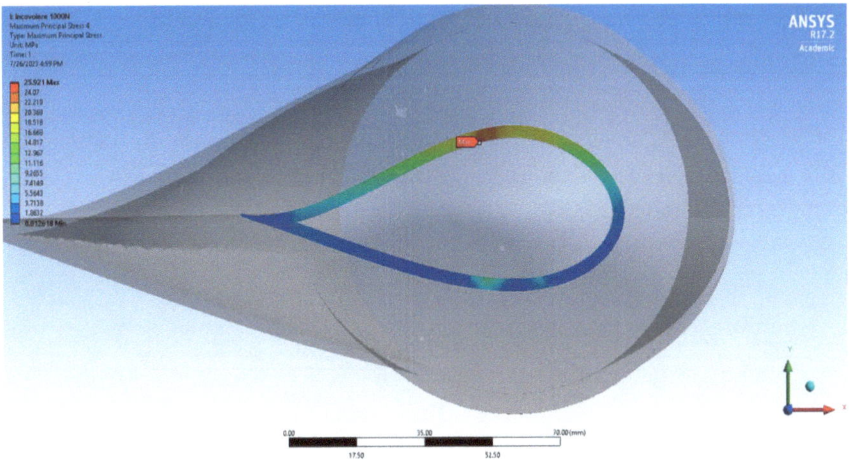

Fig. 2.13 Maximum principal stress at 130 mm from the blade tip

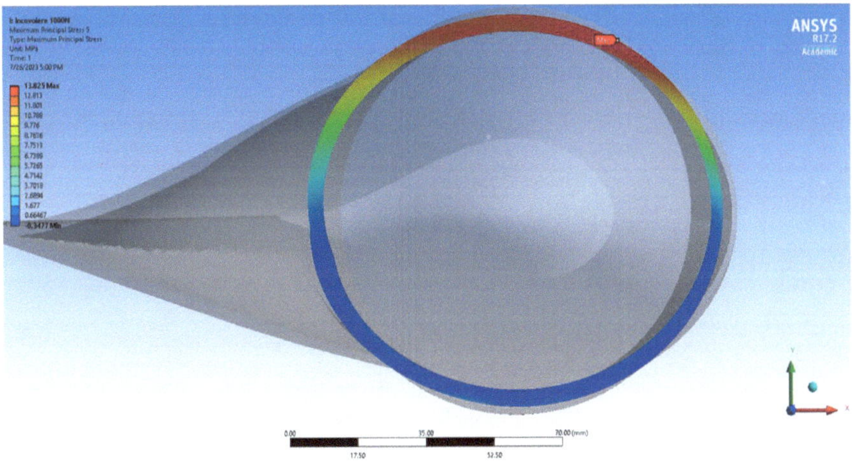

Fig. 2.14 Maximum principal stress at 230 mm from the blade hub

2.5 Case 2: Finite element analysis at clockwise torsion Mt= −170660.646 N·mm

Figure 2.15 shows the loading condition that is achieved by clamping the hub and applying a torque of M_t = -170660.646 N·mm clockwise, as viewed from the free end of the blade.

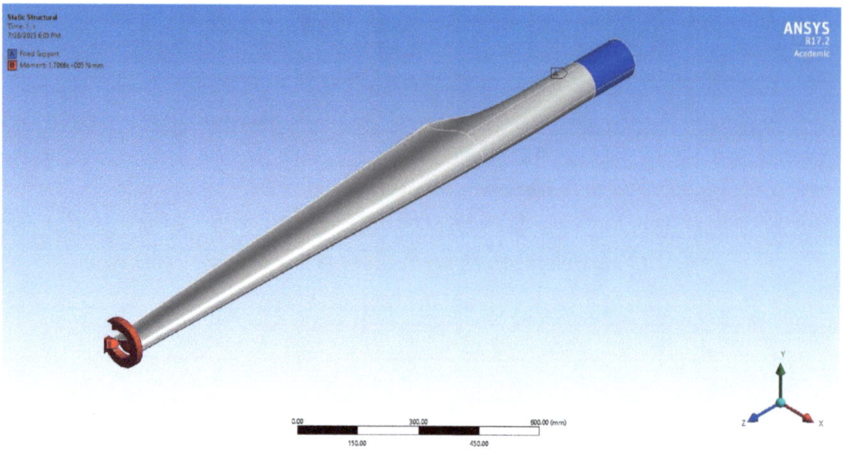

Fig. 2.15 Load condition at torsional load (clockwise)

The distributions of the von Mises stress

Figure 2.16 present the equivalent stress map with a maximum value of 49.15 MPa occurring on the inner surface of the extrados. The coordinates of the point where the maximum von Mises stress was obtained, according to the attached coordinate system, are: X: -59.99 mm; Y: -3.03 mm; Z=1730.83 mm.

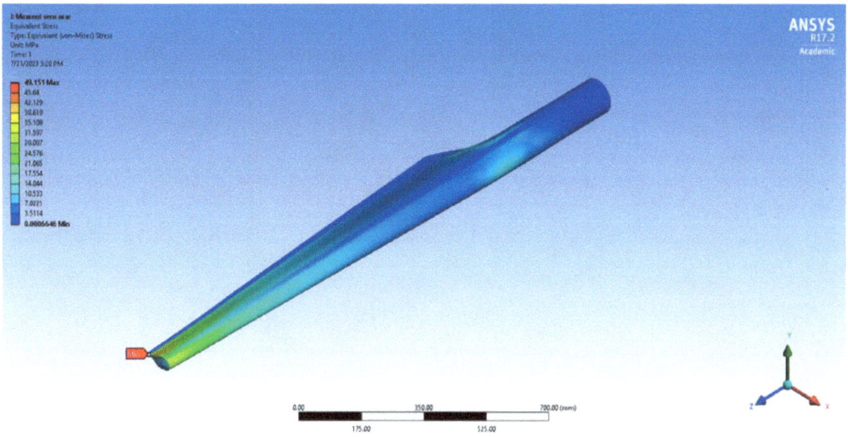

a)

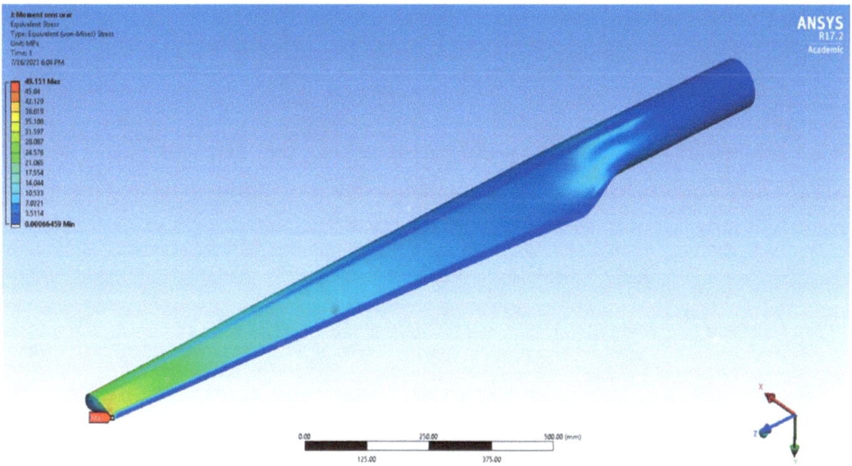

b)

Fig. 2.16 The distributions of the von Mises stress: on the intrados a) and, respectively, on the extrados b) of the blade

Figure 2.17 presents the equivalent von-Mises stress, detail from the blade tip.

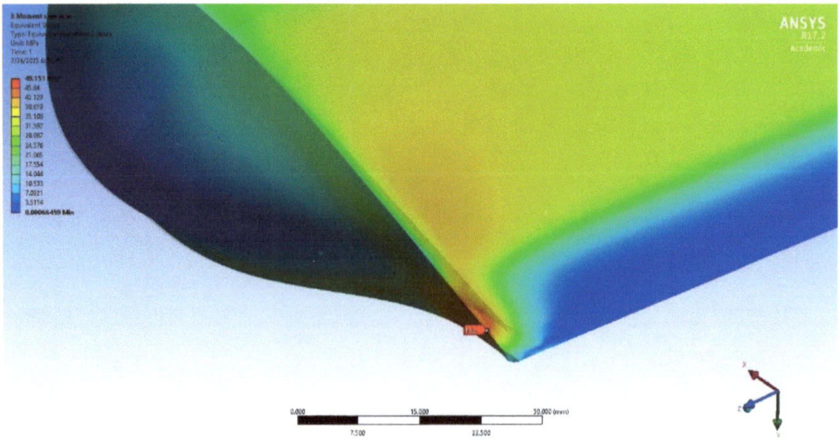

Fig. 2.17 Detail of the blade tip

Distribution of maximum principal stresses

Figure 2.18 shows the maximum principal stress distribution, $\sigma_{1\ max}$ =46.83 MPa. The coordinates of the point where the maximum principal stress occurred: X: -54.79 mm; Y: -3.74 mm; Z=1729.83 mm.

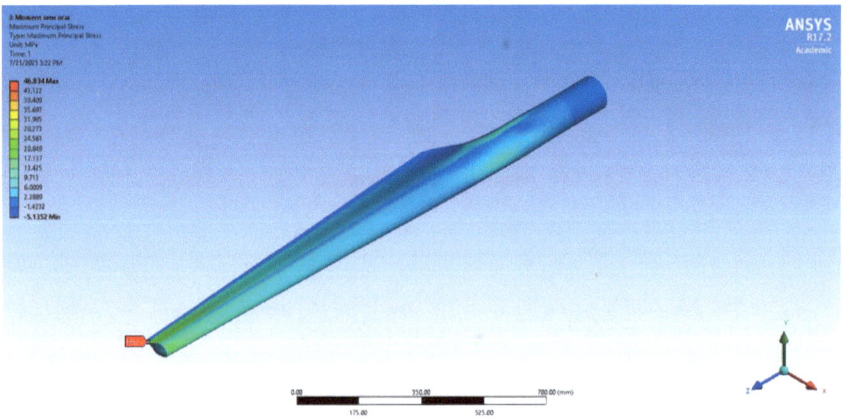

a)

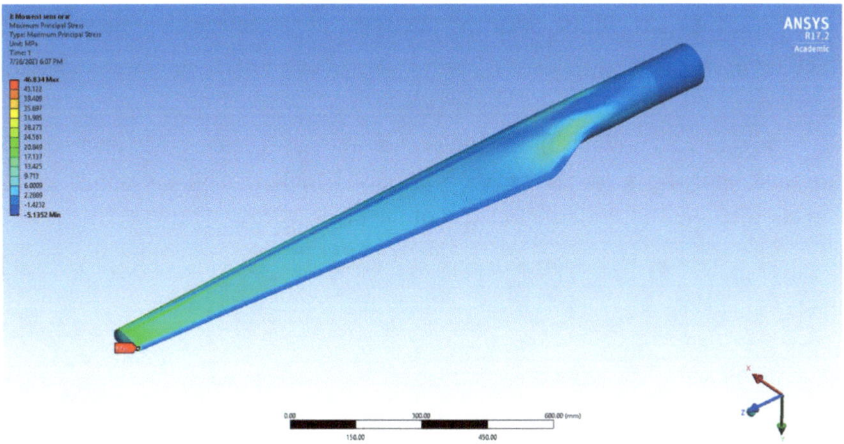

b)

Fig. 2.18 The maximum principal stress: on the intrados a) and, respectively, on the extrados b) of the blade

Figure 2.19 shows the maximum principal stress distribution - blade tip detail.

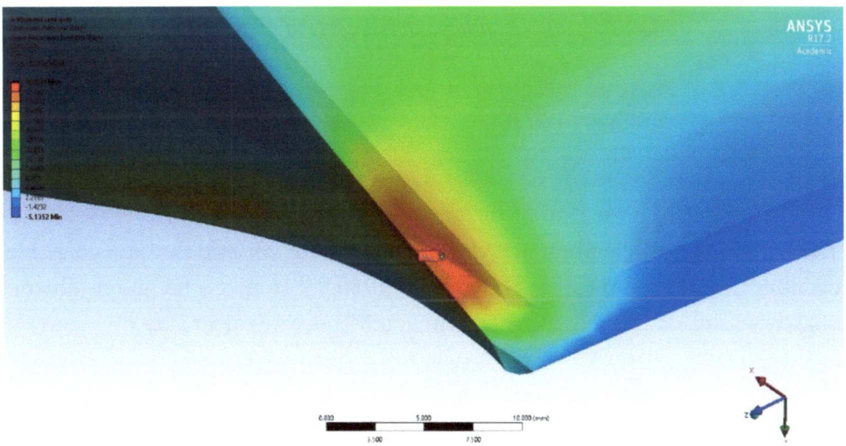

Fig. 2.19 Maximum principal stress distribution - blade tip detail

2.6 Case 3: Finite element analysis at anticlockwise torsion Mt= 170660.646 N·mm

Figure 2.20 presents the loading condition that is achieved by clamping the hub and applying a torque of Mt = 170660.646 N·mm in the anticlockwise direction, as viewed from the free end of the blade.

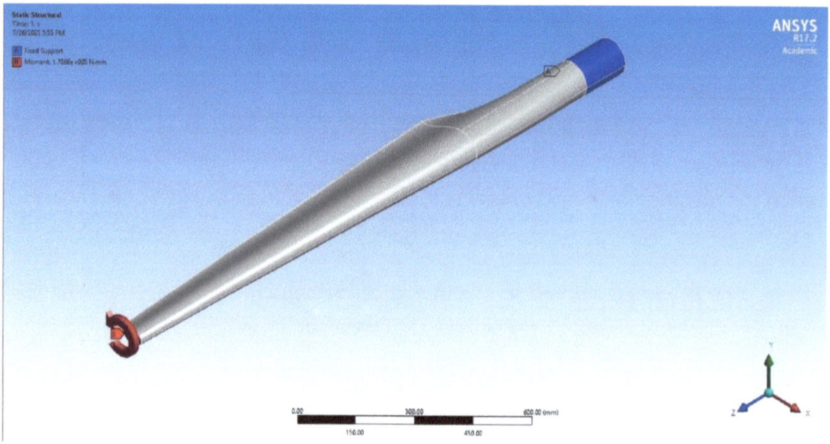

Fig. 2.20 Load condition at torsional load anticlockwise

The distributions of the von Mises stress

Figure 2.21 presents the von Mises stress distribution, $\sigma_{ech\ max} = 49.15$ MPa. The coordinates of the point at which the maximum von Mises equivalent stress occurred: X: -59.99mm; Y: -59.99; Z=1730.83 mm. Same point obtained clockwise and same value of maximum von Mises equivalent stress.

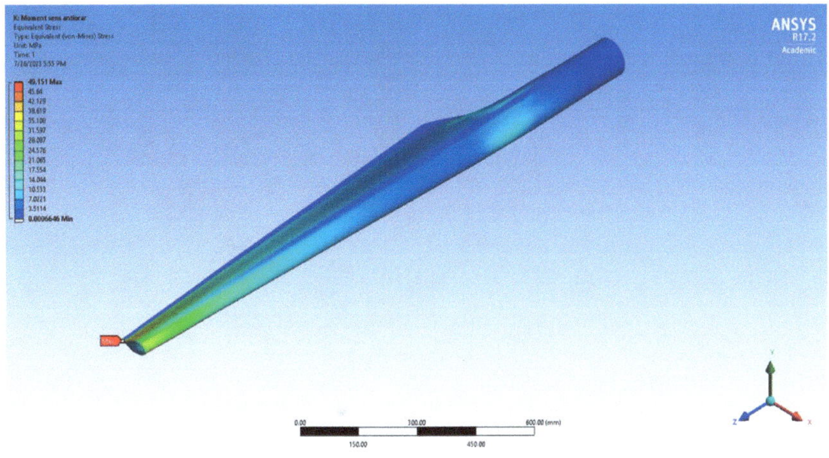

a)

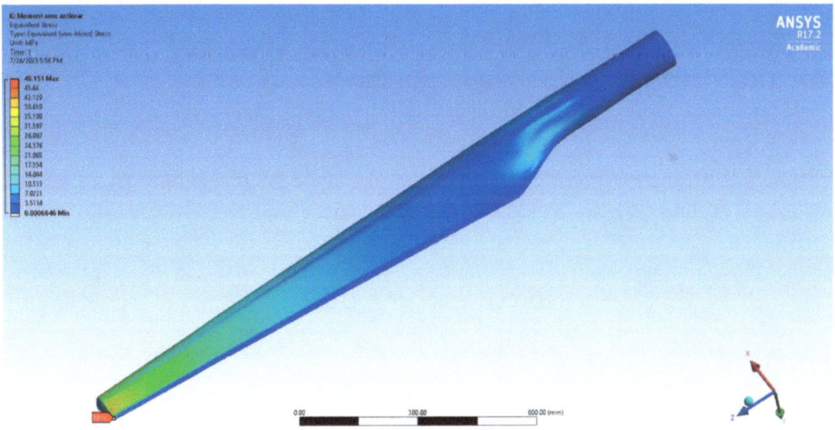

b)

Fig. 2.21 The distributions of the von Mises stress: on the intrados a) and, respectively, on the extrados b) of the blade

Figure 2.22 presents the von Mises equivalent stress, blade tip detail.

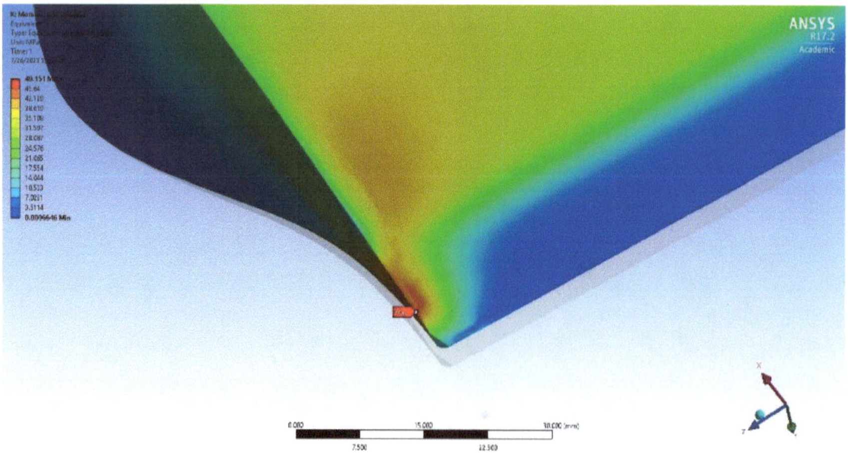

Fig. 2.22 Von-Mises equivalent stress, detail from the blade tip

Distribution of maximum principal stresses

Figure 2.23 presents the maximum principal stress distribution, $\sigma_{1\ max}$ =46.81 MPa. The coordinates of the point at which the maximum principal stress occurred: X: -54.79 mm; Y: -1.16 mm; Z=1729.83 mm.

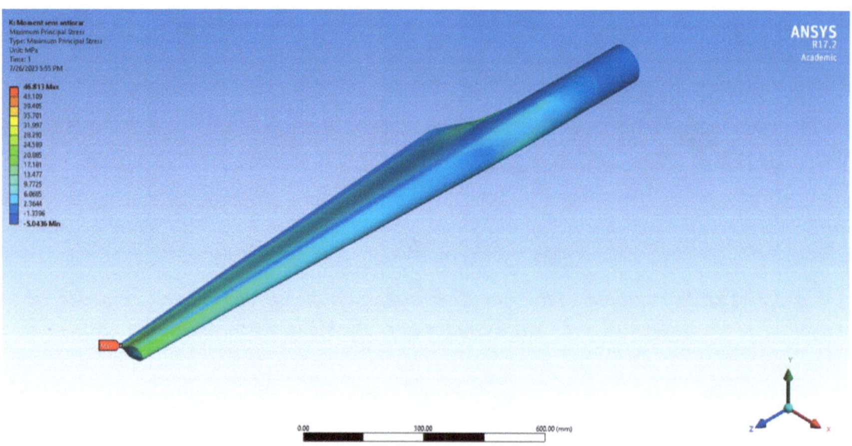

a)

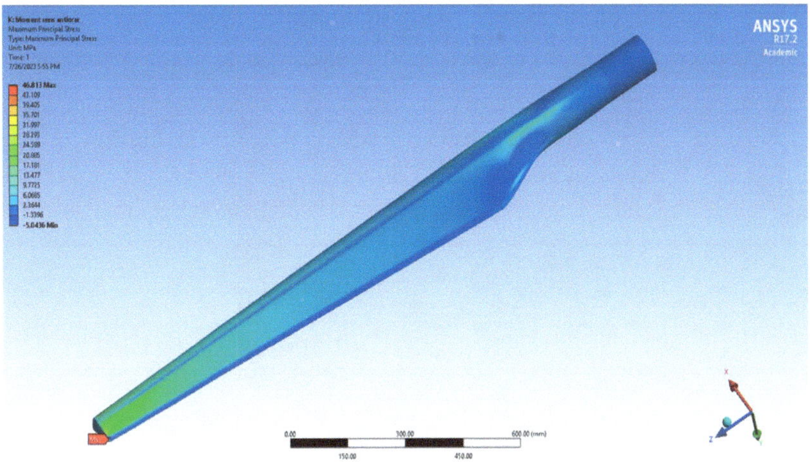

b)

Fig. 2.23 The maximum principal stress: on the intrados a) and, respectively, on the extrados b) of the blade

Figure 2.24 shows the maximum principal stress distribution - blade tip detail.

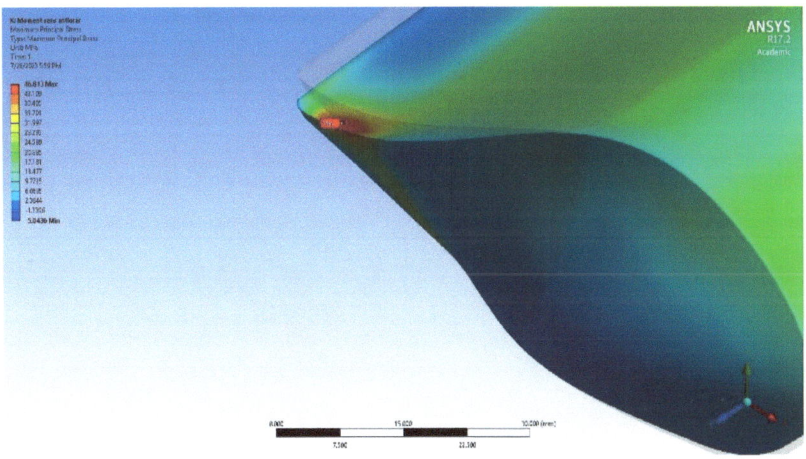

Fig. 2.24 Maximum principal stress distribution - blade tip detail.

2.7 Case 4: Finite element analysis under compound load (bending + clockwise torsion) F=1000N + Mt= −170660.646 N·mm

Figure 2.25 presents the loading condition which is realized by clamping the hub and applying a force of 1000 N in the intrados-extrados direction on the blade tip surface and a torsional moment M_t = -170660.646 in the clockwise sense.

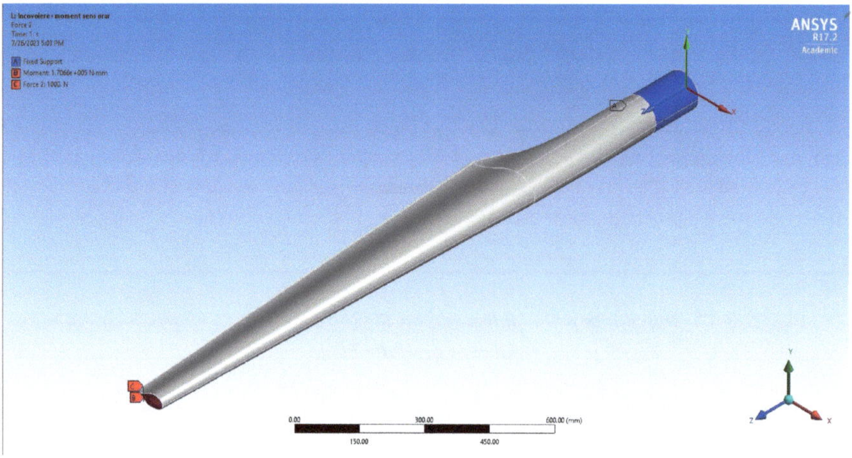

Fig. 2.25 Load condition at compound load

Figure 2.26 shows in detail, for a view from the blade tip, how to apply the force and torsional moment (red area).

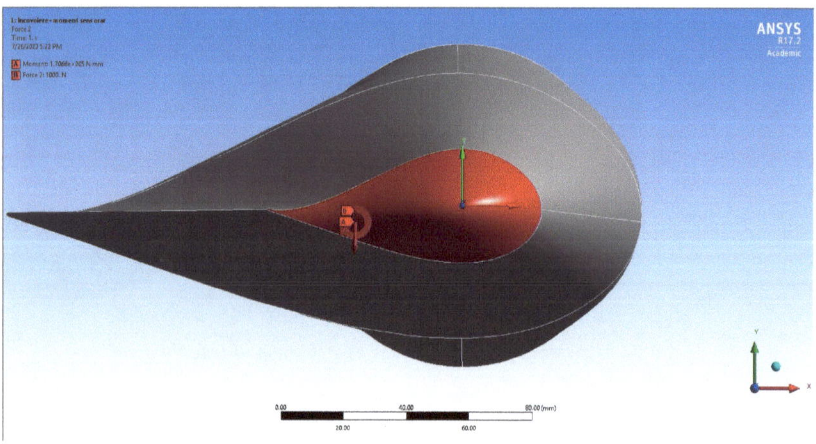

Fig. 2.26 Force and moment application detail

The distributions of the von Mises stress

Figure 2.27 shows the von Mises stress distribution, $\sigma_{ech\ max}$ =73.05 MPa. The coordinates of the point at which the maximum von Mises equivalent stress occurred: X: -6.29 mm; Y: -20.46 mm; Z=1423.93 mm.

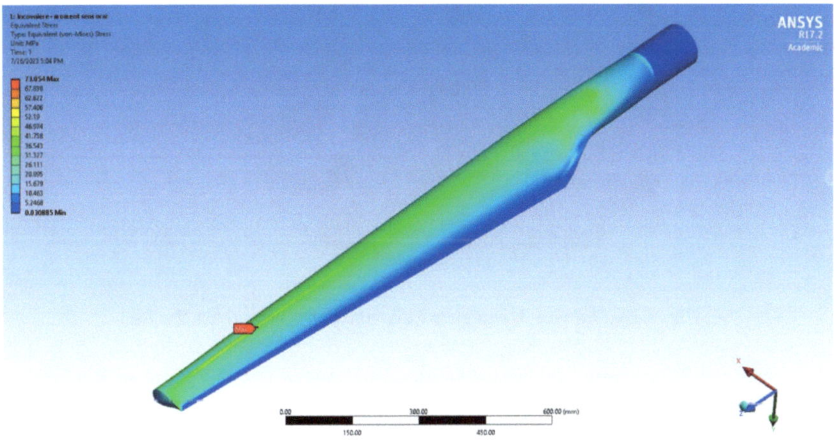

a)

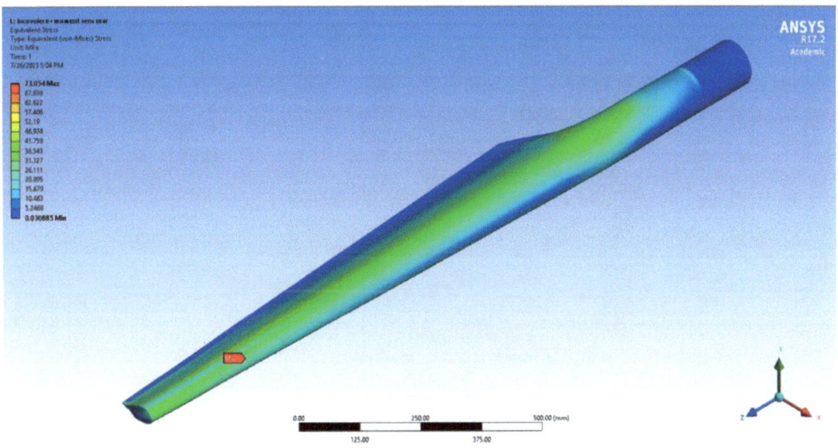

b)

Fig. 2.27 The distributions of the von Mises stress: on the intrados a) and, respectively, on the extrados b) of the blade

Figure 2.28 shows the von Mises stress distribution as seen from the hub.

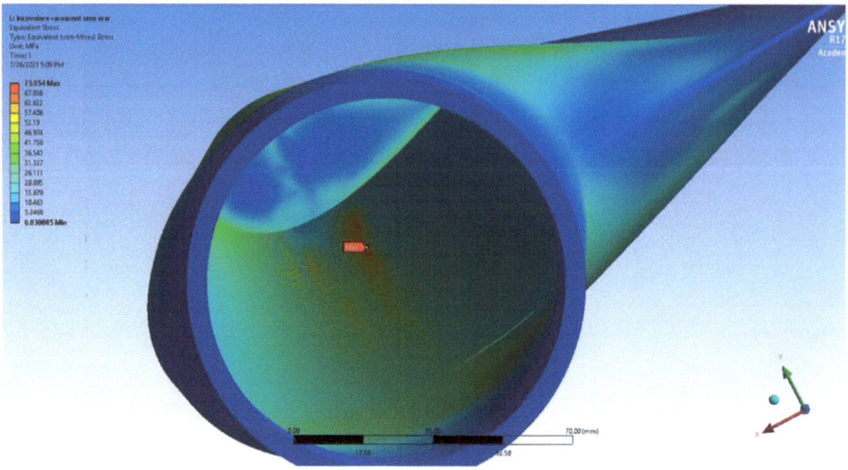

Fig. 2.28 The von Mises stress distribution - view from the hub

Figure 2.29 shows the von Mises stress distribution - view in the plane in which the maximum equivalent stress occurred.

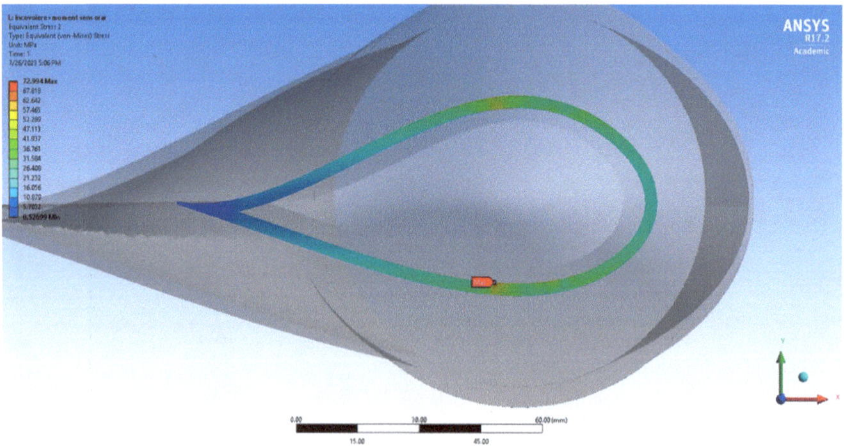

Fig. 2.29 The distribution of von Mises stresses seen in the plane in which the maximum equivalent stress occurred

Figure 2.30 presents the equivalent stress distribution at 130 mm from the blade tip, $\sigma_{ech\ max}$=52.61 MPa.

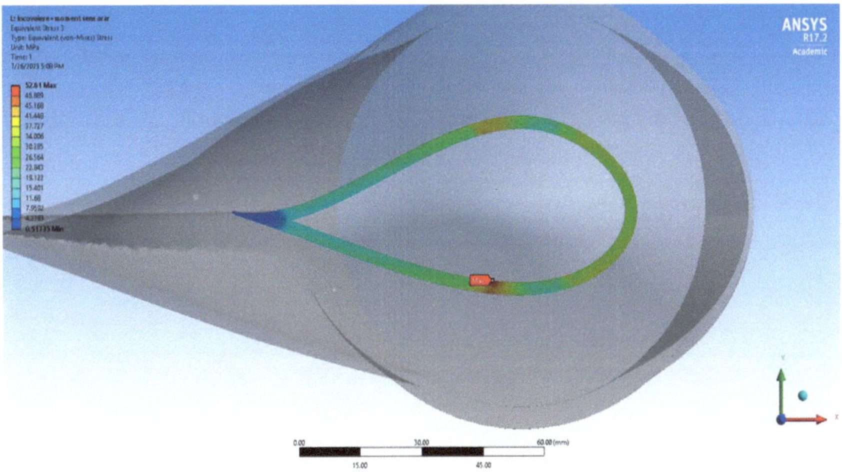

Fig. 2.30 Equivalent stress distribution at 130 mm from the blade tip

Figure 2.31 shows the equivalent stress distribution at 230 mm from the base of the blade $\sigma_{ech\ max}$=14.15 MPa.

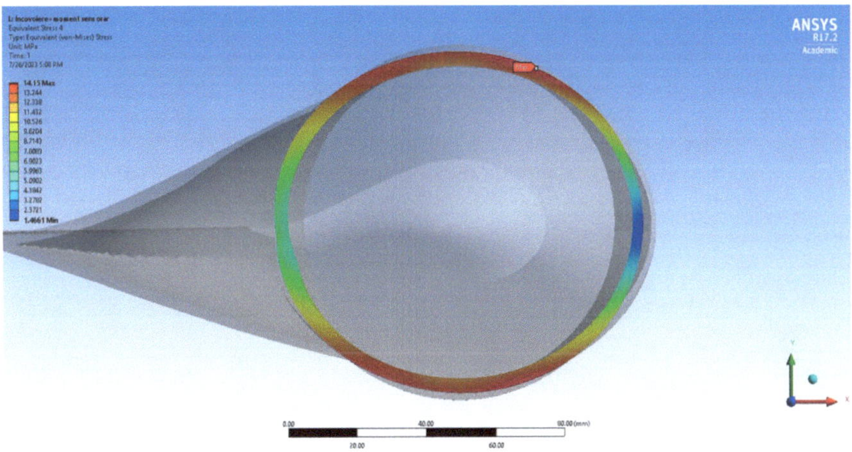

Fig. 2.31 Equivalent stress distribution at 230 mm from the blade base

Distribution of maximum principal stresses

Figure 2.32 shows the maximum principal stress distribution, $\sigma_{1\,max}$ =60.30 MPa. The coordinates of the point at which the maximum principal stress occurred: X: -7.42 mm; Y: 18.43 mm; Z=1507.13 mm.

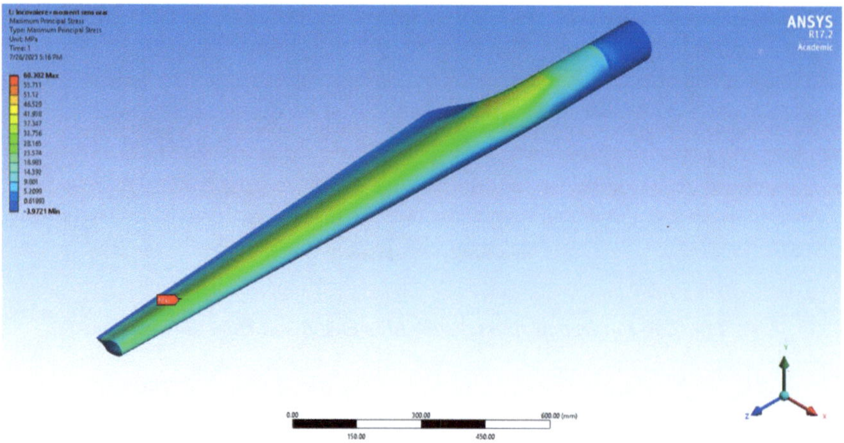

a)

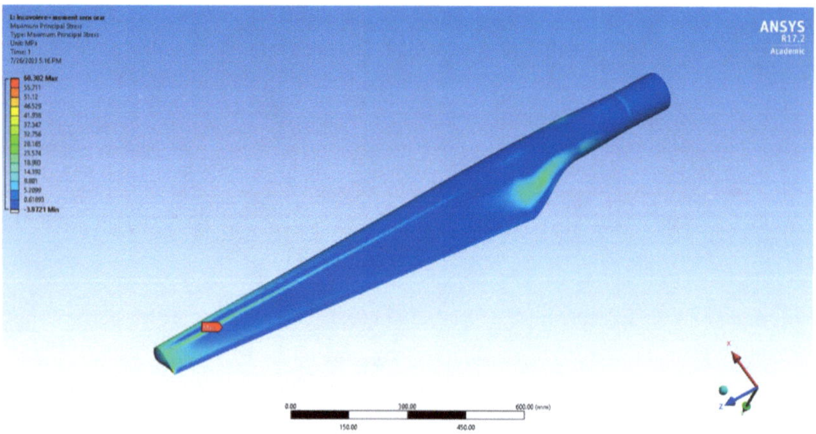

b)

Fig. 2.32 The maximum principal stress: on the intrados a) and, respectively, on the extrados b) of the blade

Figure 2.33 presents the view in the plane where the maximum principal stress occurred.

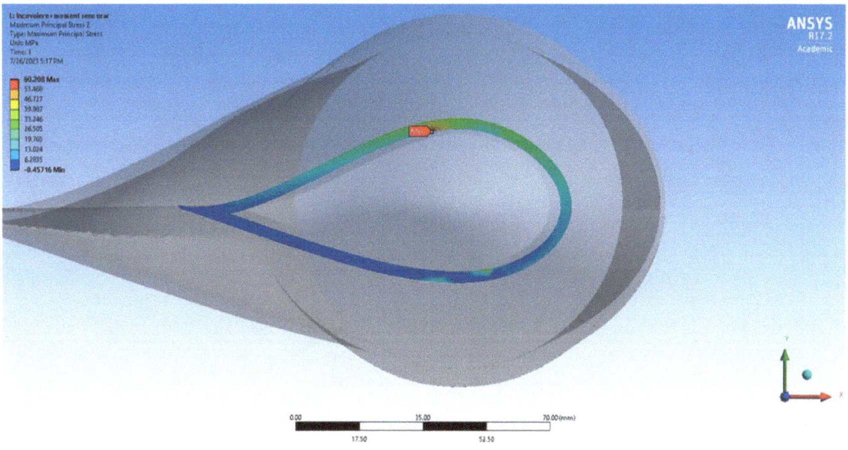

Fig. 2.33 View in the plane where the maximum principal stress occurred

Figure 2.34 shows the principal stress distribution at 130 mm from the blade tip, $\sigma_{1\,max} = 46.43$ MPa.

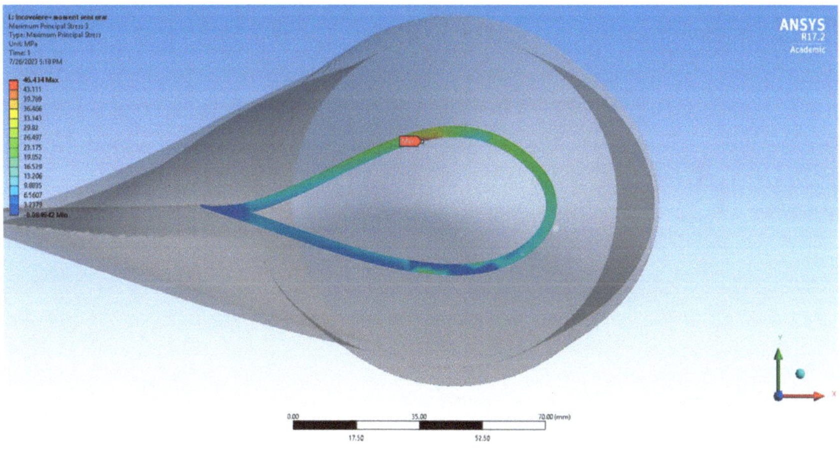

Fig. 2.34 Principal stress distribution at 130 mm from the blade tip

Figure 2.35 shows the principal stress distribution at 230 mm from the base of the blade, $\sigma_{1\,max}$=13.71 MPa.

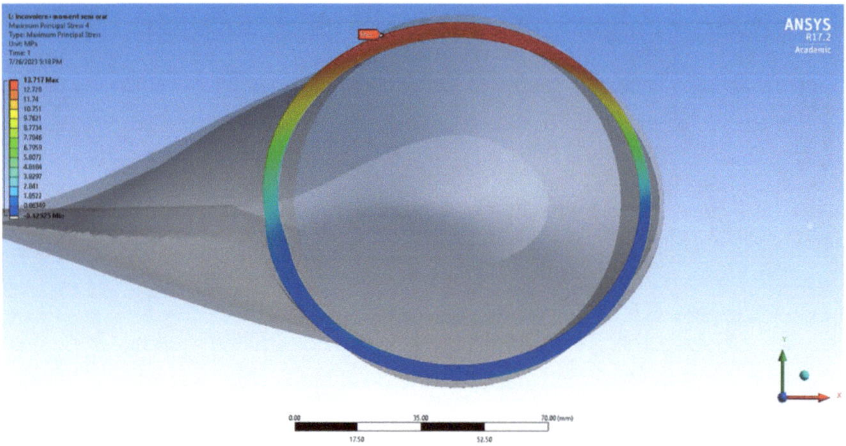

Fig. 2.35 Principal stress distribution at 230 mm from the blade base

2.8 Case 5: Finite element analysis under compound load (Bending + Anticlockwise torsion F=-1000N + Mt= 170660.646 N·mm)

Figure 2.36 presents the loading condition which is achieved by clamping the hub and applying a force of 1000 N in the intrados-extrados direction on the blade tip surface and a torsional moment M_t = 170660.646 N·mm in the anticlockwise sense.

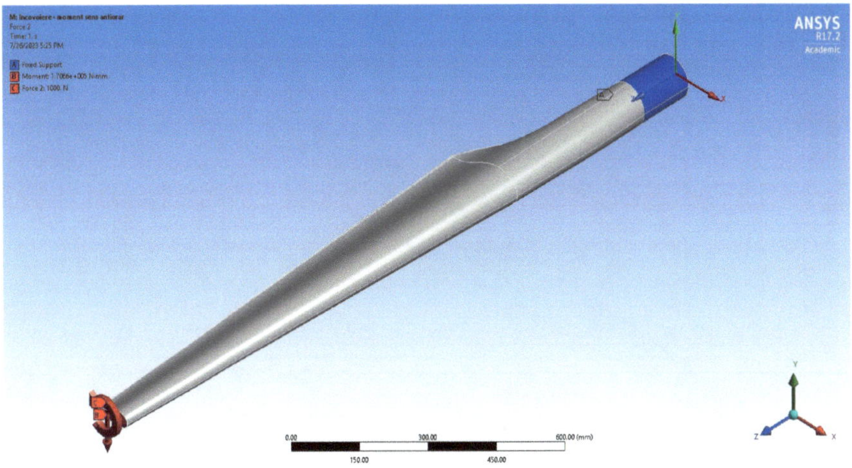

Fig. 2.36 Load condition at compound load

The distributions of the von Mises stress

Figure 2.37 presents the von Mises stress distribution, $\sigma_{ech\ max}$ =67.07 MPa. The coordinates of the point at which the maximum von Mises equivalent stress occurred: X: -6.29 mm; Y: -20.46 mm; Z=1423.93 mm. In the compound numerical simulations (bending + torsion), the maximum equivalent stress appears at the same point, but with different values.

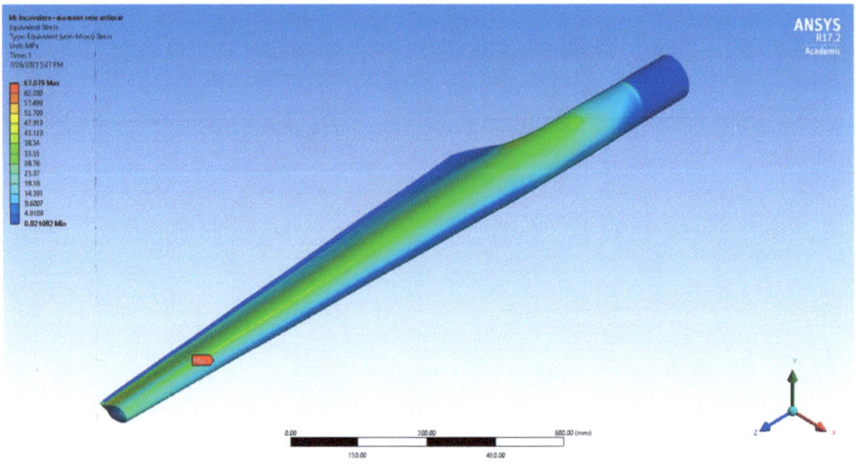

a)

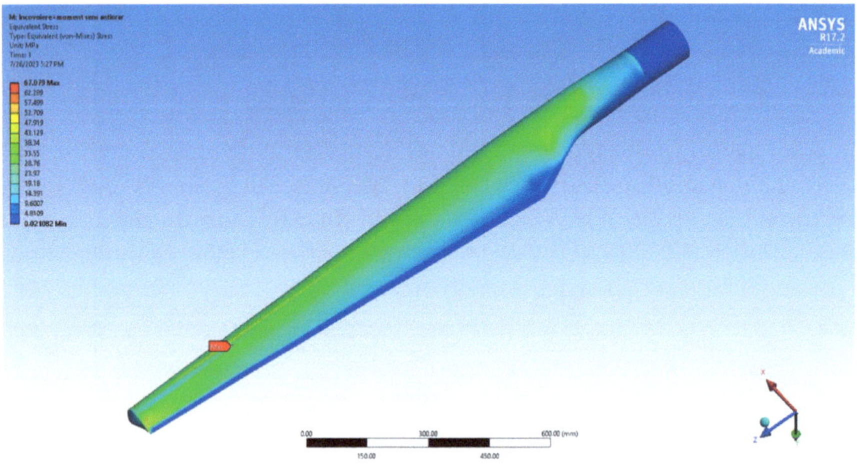

b)

Fig. 2.37 The distributions of the von Mises stress: on the intrados a) and, respectively, on the extrados b) of the blade

Figure 2.38 illustrates the view in the plane where the maximum equivalent stress occurred.

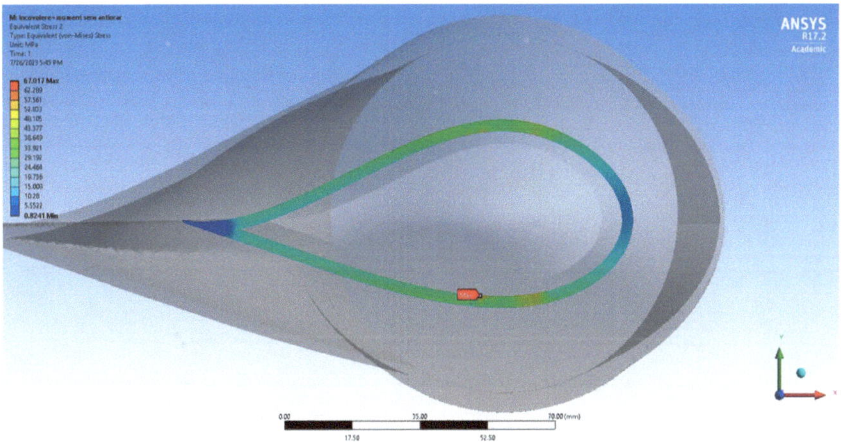

Fig. 2.38 View in the plane where the maximum equivalent stress occurred

Figure 2.39 presents the maximum principal stress in the plane where the maximum equivalent stress occurred.

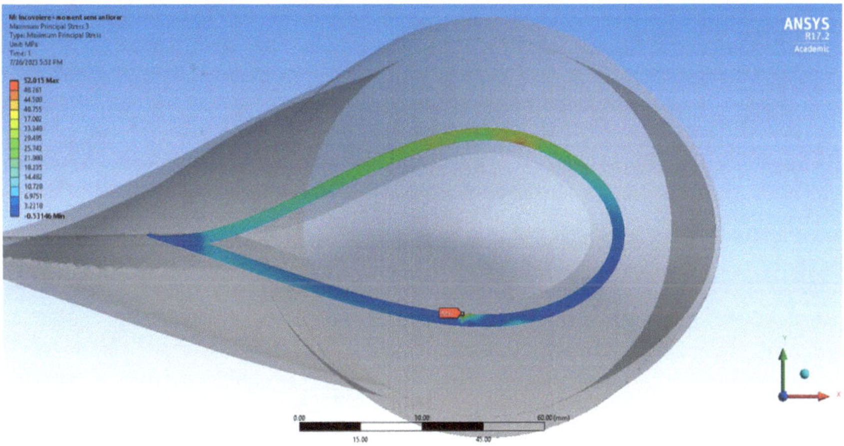

Fig. 2.39 View in the plane where the maximum principal stress occurred

Distribution of maximum principal stresses

Figure 2.40 presents the principal stress distribution at 130 mm from the blade tip, $\sigma_{ech\ max}$ = 41.02 MPa.

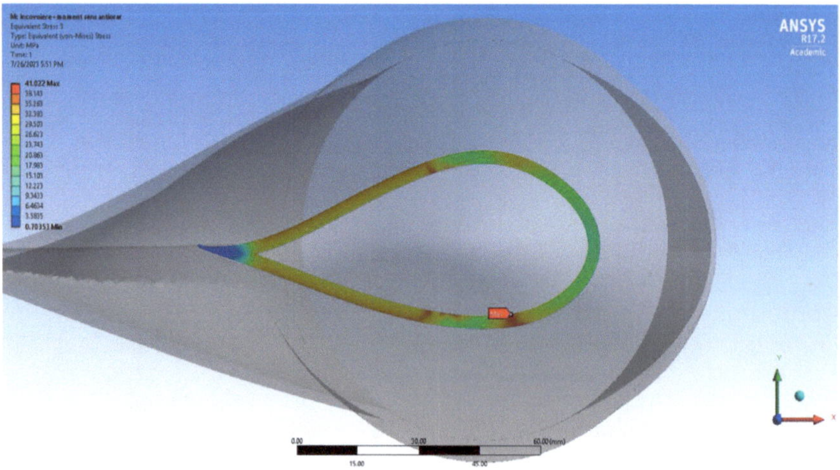

Fig. 2.40 Principal stress distribution at 130 mm from the blade tip

Figure 2.41 illustrates the principal stress distribution at 230 mm from the base of the blade, $\sigma_{ech\ max}$=14.49 MPa.

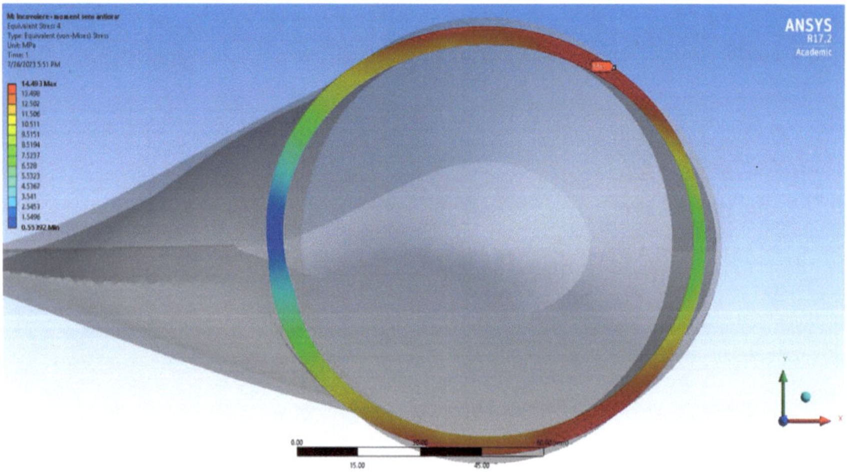

Fig. 2.41 Principal stress distribution at 230 mm from the blade base

Distribution of maximum principal stresses

Figure 2.42 presents the maximum principal stress distribution, $\sigma_{1\ max}$ =52.07 MPa. The coordinates of the point at which the maximum principal stress occurred: X: -6.29 mm; Y: -20.46 mm; Z=1423.93 mm.

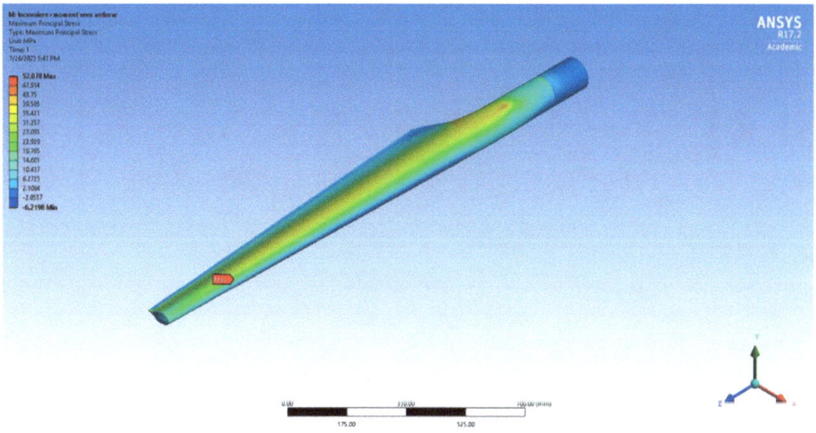

a)

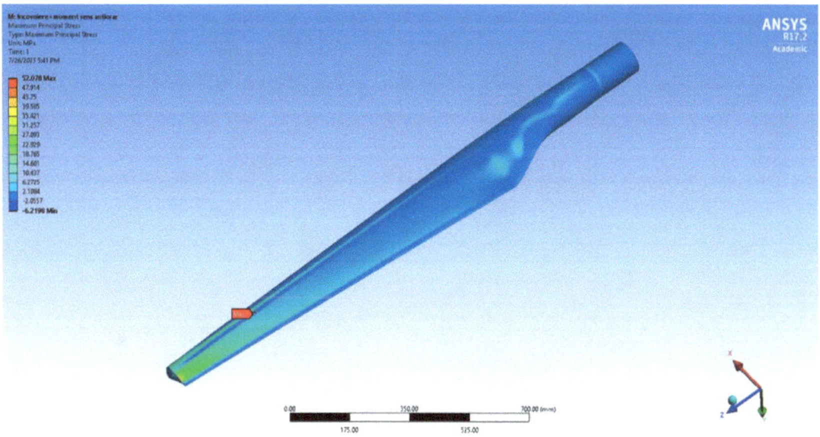

b)

Fig. 2.42 The maximum principal stress: on the intrados a) and, respectively, on the extrados b) of the blade

Figure 2.43 illustrates the view in the plane in which the maximum principal stress occurred.

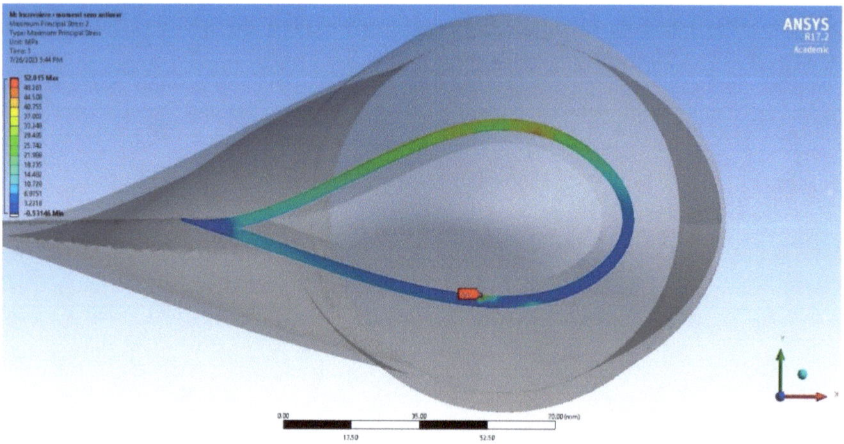

Fig. 2.43 View in the plane where the maximum principal stress occurred

Figure 2.44 presents the distribution of principal stresses at 130 mm from the blade tip, $\sigma_{1\,max}$ = 34.75 MPa.

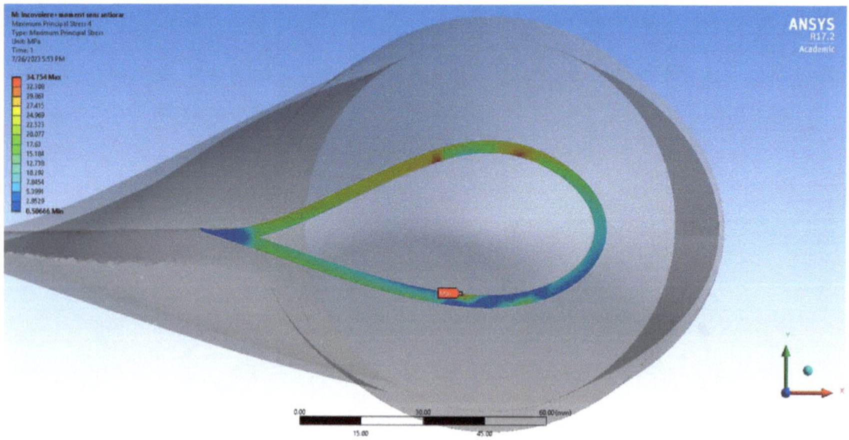

Fig. 2.44 Distribution of principal stresses at 130 mm from the blade tip

Figure 2.45 presents the distribution of principal stresses at 230 mm from the base of the blade, $\sigma_{1\ max}$=14.63 MPa.

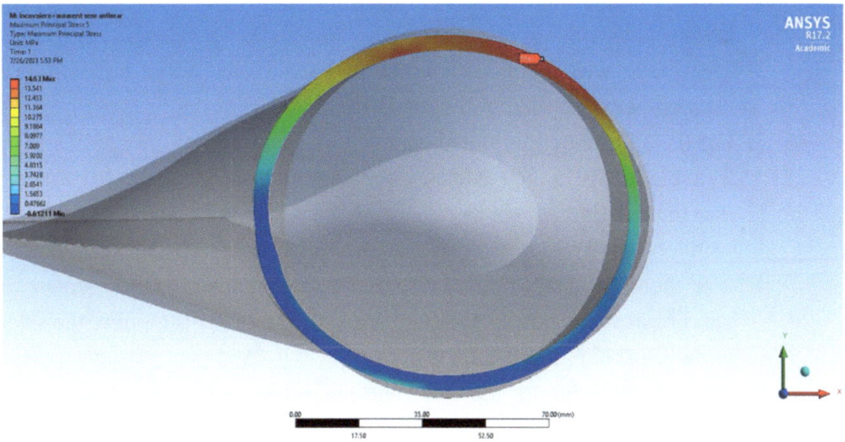

Fig. 2.45 Distribution of principal stresses at 230 mm from the blade base

In Table 4.1, both the distribution of the von Mises equivalent stresses and the maximum principal stress for the five stress cases have been centralized.

Tabel 4.1 Distribution of von Mises equivalent stress and maximum principal stress

Nt.crt	**Case study**	$\sigma_{ech\ max}$	$\sigma_{1\ max}$
1	Bending F=1000N	**60.318 MPa** X: -6.282 mm, Y: -14.13 mm, Z:1425.93 mm	**41.83 MPa** X: -5.729 mm, Y: 23.85 mm, Z: 1254.23 mm
2	Clockwise torsion M_t= −170660.646 Nmm	**49.15 MPa** X: -59.99 mm; Y: -3.03 mm; Z=1730.83 mm	**46.83 MPa** X: -54.79 mm; Y: -3.74 mm; Z=1729.83 mm
3	Anticlockwise torsion M_t=170660.646 Nmm	**49.15 MPa** X: -59.99 mm; Y: -3.03 mm; Z=1730.83 mm	**46.81 MPa** X: -54.79 mm; Y: -1.16 mm; Z=1729.83 mm
4	Bending + clockwise torsion F=1000 N + M_t= −170660.646 Nmm	**73,05 MPa** X: -6.29 mm; Y: -20.46 mm; Z=1423.93 mm	**60.30 MPa** X: -7.42 mm; Y: 18.43 mm; Z=1507.13 mm
5	Bending + Anticlockwise torsion F=1000 N + M_t= 170660.646 Nmm	**67.07 MPa** X: -6.29mm; Y: -20.46mm; Z=1423.93mm	**52.07 MPa** X: -6.29 mm; Y: -20.46 mm; Z=1423.93 mm

2.9. General conclusions to Chapter 2

1. Finite element analysis of wind turbine blade was carried out on five cases;
2. The blade was embedded towards the hub and force and torsional moment were applied to the blade tip;
3. For the simple bending load case with a force of 1000 N applied at the free blade tip, a von Mises equivalent stress of 60.31 MPa and a maximum principal stress of 41.83 MPa occurring at the blade intrados were obtained;
4. For the case of both clockwise and anticlockwise twisting stress, the maximum stress values obtained are approximately the same, the stress variations being similar;
5. The highest value for the von Mises equivalent stress is obtained in the case of the compound bending and twisting stress clockwise of 73.05 MPa, where the coordinates of the point where the maximum von Mises equivalent stress occurred were X: -6.29 mm; Y: -20.46 mm; Z=1423.93 mm. At the same coordinates, the von Mises equivalent stress also appears, but in anticlockwise direction;
6. The von Mises equivalent stress in the case of compound bending and twisting stress in counter-clockwise direction is 67.07 MPa.

CHAPTER 3. TESTING A SMALL-SCALE BLADE UNDER COMPOUND LOAD CONDITIONS

3.1 The purpose of testing the wind turbine blade at scale

The aim of testing a wind turbine blade at 1:10 scale is to observe its behavior under varying stress conditions (simple and compound stresses).

The test control parameters were: torque M_t, angle of rotation, force and displacement. During the test, signals were acquired from the sensors used: displacement sensors, strain gauge rectangular and chain strain gauges with delta transducers, force cells. By processing and interpreting the experimental data, computational algorithms were developed and they identified critical areas on the wind turbine blade. Maximum values of some quantities have been determined to estimate where blade failure will occur.

The knowledge gained from this testing process is directly applicable to the development of real-time structural health monitoring (SHM) systems for wind turbine blades. By integrating these results into SHM platforms, operators can continuously assess the integrity of blades during operation. Such systems would enable early detection of damage, increase operational safety, and minimize the likelihood of catastrophic failure. Furthermore, predictive maintenance strategies based on these monitoring systems would lower overall maintenance costs, optimize inspection schedules, and significantly reduce turbine downtime for repairs, thereby improving the economic efficiency of wind energy production.

3.2 The experimental set-up and sensors used

The tests were performed on a 1:10 scale wind turbine blade made of GFRP. In order to have both simple and complex blade stresses, an experimental stand was made that individually and simultaneously applies torsional and bending stress.

Four strain gauge rosettes, type C2A-06-062WW-350, manufactured by Micro-Measurements, were mounted on the blade with pre-attached cables, each with three grids, as in Figure 3.1. The characteristics of the strain gauge used are as follows:

- Resistance: 350Ω±0.6%;
- Gauge factor: k_G=2,16±0.5%;
- Transverse sensibility K_t= +0.8%;
- Support material: polyimide;
- Grid material: constantan.

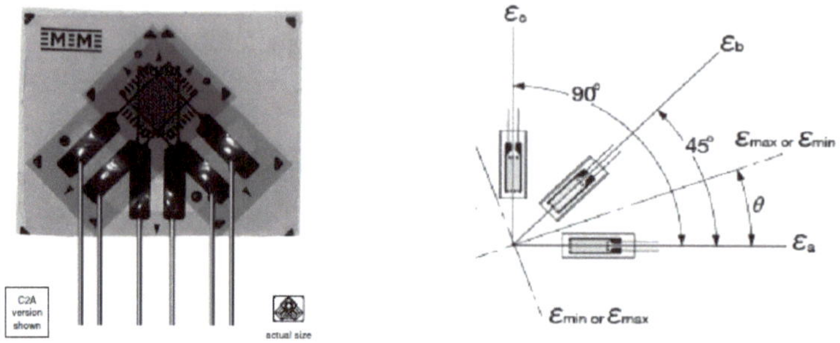

a) *b)*

Fig.3.1 Strain gauge rosettes (a); notations used (b)

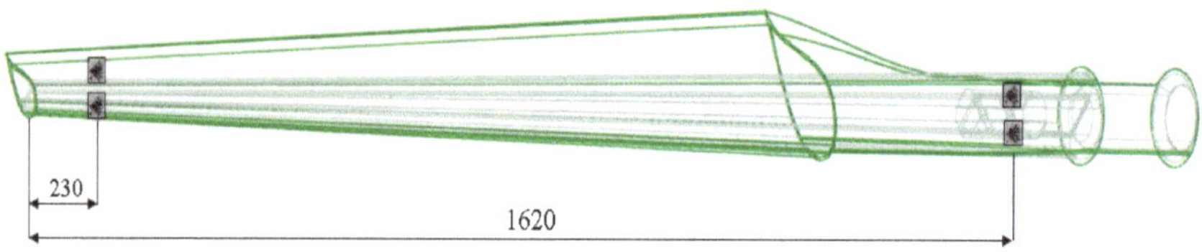

Fig. 3.2 Overview of the location of rectangular strain gauges

The strain gauges were applied in two sections: the section located 1620 mm from the end of the grip beyond the grip hub area (transducers T1 and T2) and the section located 230 mm from the blade tip (T3, T4), as shown in Figure 3.2. Figure 3.3 presents, in detail and in cross-section, how the four strain gauges are placed.

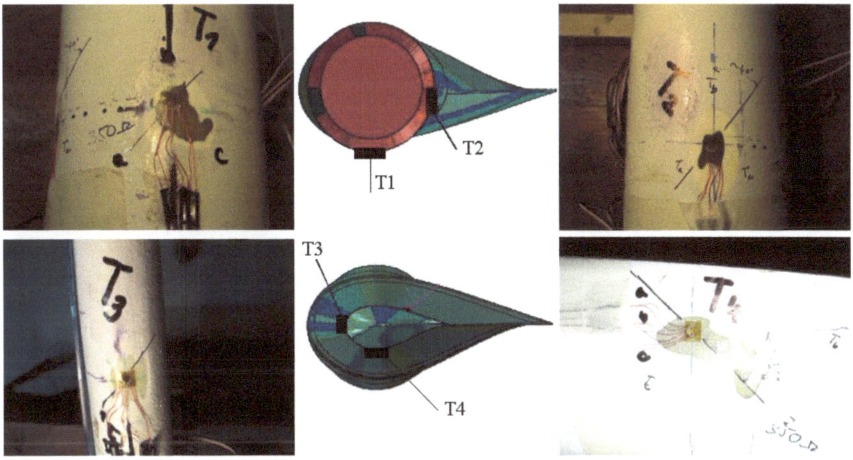

Fig. 3.3 Location details of T1, T2, T3 and T4 strain gauges on the blade

The acquisition of the strain from the four strain gauges was carried out with Vishay P3 tensor bridges, as shown in Figure 3.4. Twelve acquisition channels were required and the type of mounting used was quarter-bridge.

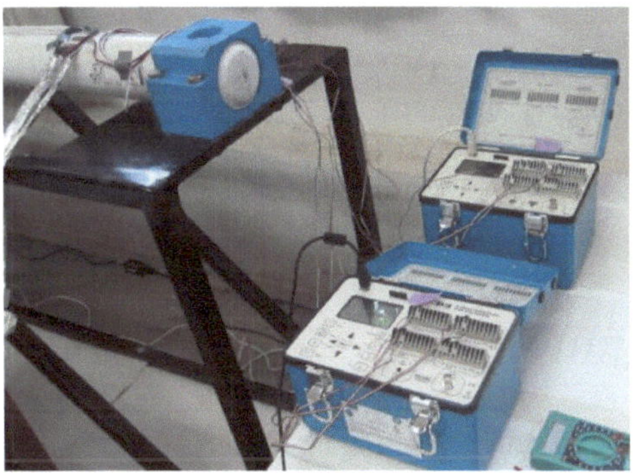

Fig. 3. 4 Vishay P3 tensor bridges

Four additional 4/120 KY31 type 4/120 strain gauge chains manufactured by HBM, as shown in Figure 3.5, each with five rosettes of three grids each positioned at 0°/120°/240°, were mounted on the blade in areas of interest identified by FEA. The characteristics of the strain gauges chains used are as follows:

- Resistance: 120Ω ± 0.50%;

- Gauge factor (a,b,c): k_G=1.83±1.0%; k_G=1.87±1.0%; k_G=1.87±1.0%;
- Transverse sensibility (a,b,c): K_t= 0.3%; K_t= 2.3%; K_t= 2.3%;
- Support material: polyimide;
- Grid material: constantan.

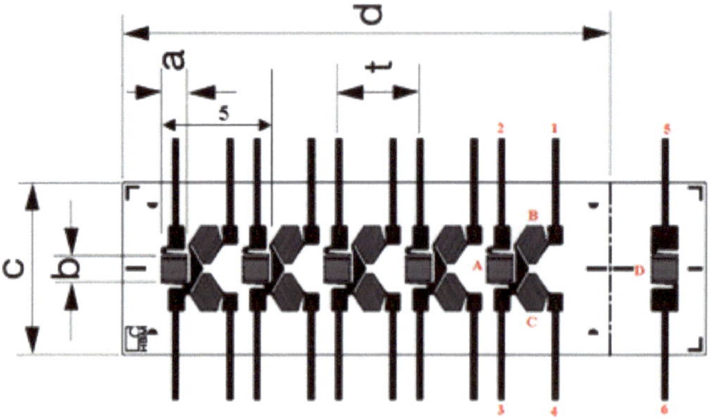

Fig. 3.5 Tensiometer chain with 5 delta strain gauges, grid and filament notation

The strain gauge chains were applied in two sections: chains 1, 2 and 3 were placed on the intrados of the blade and chain 4 on the extrados of the blade, as shown in Figure 3.6.

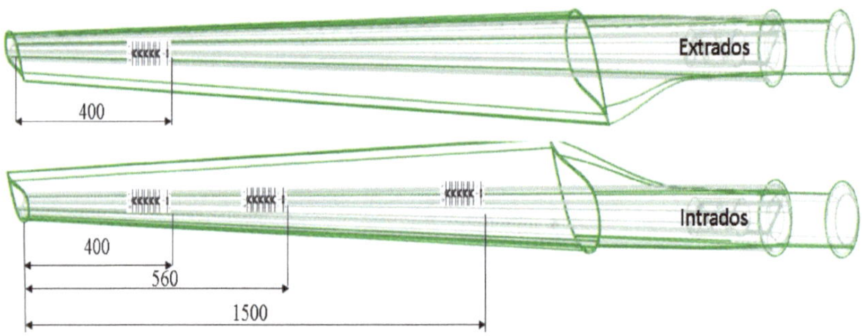

Fig. 3.6 Overview of extrados and intrados placement

Figure 3.7 presents, in detail, how to position the 4 strain gauge chains.

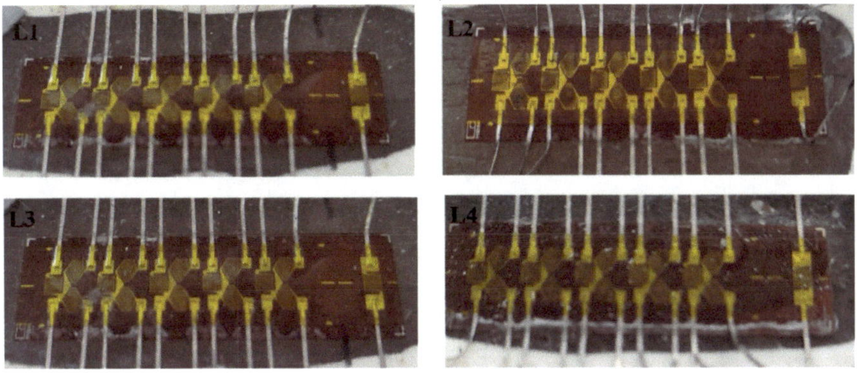

Fig. 3.7 Details of the location of the strain gauge chains on the blade

The acquisition of deformations from the 4 chains (20 electrical transducers x 3 strain gauges on each transducer) was performed with 4 QUANTUM HBM MX1615B data acquisition systems, as shown in Figure 3.8.

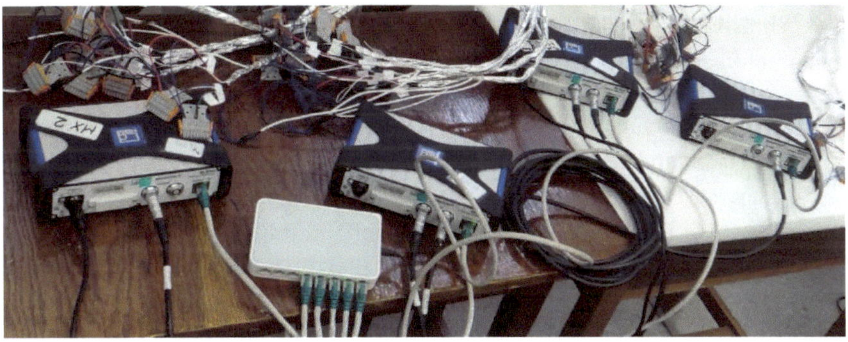

Fig. 3.8 QUANTUM HBM MX1615B acquisition system

The measurements were performed in dynamic mode, with a sampling frequency of 20 Hz, and the connection in the software interface was of the "sg 3 wire 120" type. The deformations were recorded using connection cables and a laptop with a Catman – AP data acquisition system, as shown in Figure 3.9.

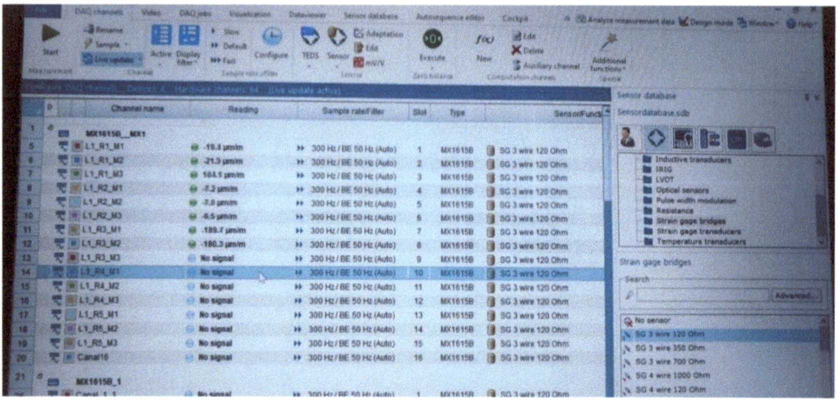

Fig. 3.9 Catman data acquisition system interface – AP

The manufacturer's instructions were followed for the application of the transducers and strain gauge chains. The blade surface was processed, removing the paint layer applied during manufacturing on the GFRP composite structure, with abrasive paper with a grain size of P100 (401). This process was carried out without affecting the fiberglass reinforcement.

After removing the resulting dust, the surface was wiped with a cotton cloth soaked in isopropyl alcohol. The rosettes and chains were glued using HBM Z-70 (cyanoacrylate) adhesive. The transducers and tinned connections were insulated from the corrosive action of the environment with liquid-applied polyurethane, which, after hardening, formed a protective film. The cables were shielded with aluminum foil, so as not to be influenced by electromagnetic fields in the environment.

The rosette grids of type C2A-06-062WW-350, when glued to the blade, were oriented as follows: the rosette grid b was oriented along the blade axis, and the grids a and c, at 90° to each other, as in Figure 3.1. With the help of these rosettes, the principal stresses and their directions from their center were determined.

The 4/120 KY31 type rosette chains, when glued to the blade, were oriented as follows: grid A parallel to the blade axis, and grids C and B oriented at 120° to each other, as in Figure 3.5. With the help of these rosette chains, the principal stresses and their directions from their centers, as well as their variation (gradient) along the length of the chain, were determined.

Figure 3.10 shows the experimental stand during testing and taking over deformations.

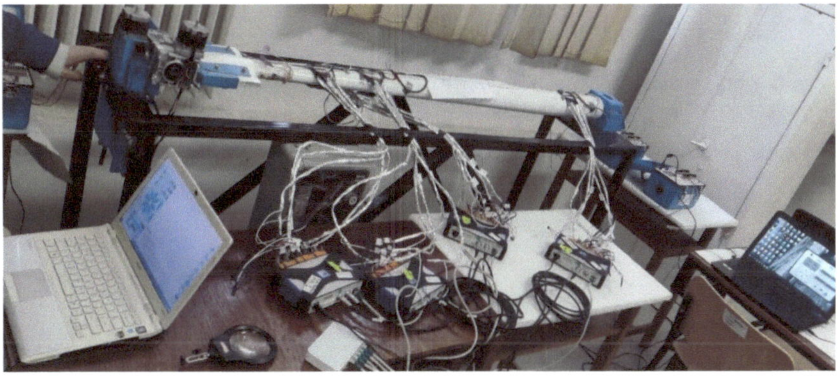

Fig. 3.10 The experimental stand during wind turbine blade testing

3.3 Electrical resistivity tensometry

Electrical resistivity tensometry is based on the phenomenon of the change in electrical resistance of conductors when they are subjected to tension or compression. The basic relationship of resistive electrical strain gage is:

$$\frac{\Delta R}{R} = k\varepsilon$$

where:

- R is the initial electrical resistance of the conductor (transducer), in a non-tensioned state;
- ΔR represents the variation in the resistance of the transducer (strain gauge), produced as a result of its tensioning;
- ε is the average specific elongation under the transducer grid;
- k is the gauge factor, indicated by the manufacturer.

Since the variation of resistance ΔR is very small, about 0.00024 Ω/με, to increase the accuracy of the measurements, the strain gauge marks are mounted in a Wheatstone bridge. The mounting of the transducers in the bridge can be done in one of the following ways:

- in quarter-bridge, as shown in Figure 3.11a, when a strain gauge is mounted in only one of the four deck arms;
- in the half-bridge, as shown in Figure 3.11b, when strain gauge marks are mounted in two arms;
- in a full bridge, as shown in Figure 3.11c, in which case a strain gauge mark is mounted in each bridge arm.

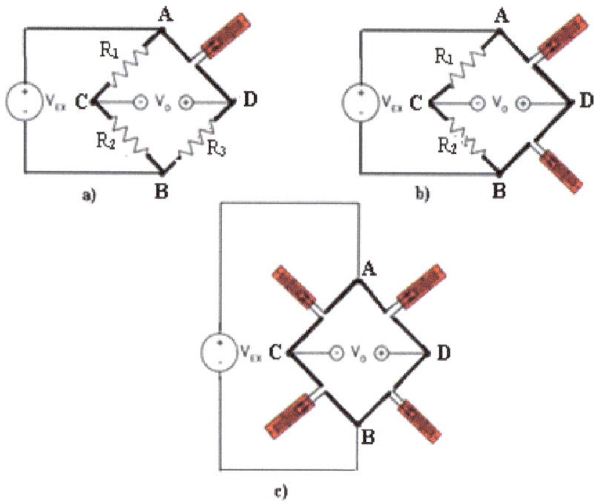

Fig. 3.11 Mounting of transducers in the bridge: a) quarter; b) half; c) full

Quarter-bridge mounting has the advantage of simplicity, low cost, but can only compensate for errors introduced by temperature variation during measurements over a limited temperature range (usually 10-15°C).

Half-bridge mounting can be realized in several variants:

- with an active and a compensating strain gauge;
- with two active strain gauges mounted in adjacent arms;
- with two active strain gauges mounted in opposing arms.

This setup has the advantage that it can compensate for errors introduced by temperature variation over a much wider range.

Full-bridge mounting has the advantage of better compensation for errors introduced by temperature variation, as well as four times the sensitivity of quarter-deck mounting, but it is expensive and more difficult to obtain because two tensile and two compressive tensile strain gauges must be tensile and two compressive to balance the bridge. This mounting is generally reserved for sensors (such as force sensors, moment sensors, etc.).

Due to the large number of transducers and cables, the quarter-bridge was chosen for practical reasons. Although this arrangement does not compensate for the errors introduced by temperature variations during the measurements over a limited temperature range, the accuracy of the experimental results was not significantly affected for the following reasons, as shown in Figure 3.12:

- the determinations were made at a temperature of about 20°C, the reference temperature for determining such errors;
- during the tests the temperature varied by only ±1°C.

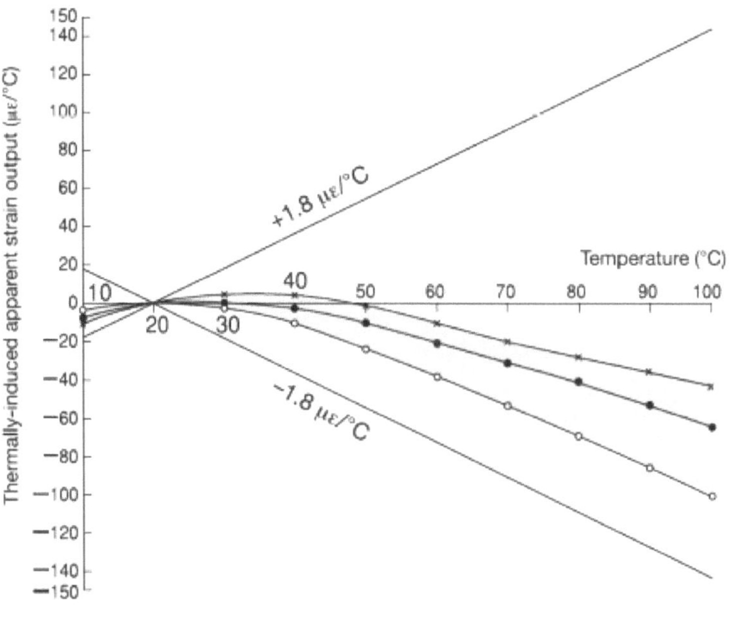

Fig. 3.12 The apparent signal induced by the temperature variation during the measurements (the gray area includes the experimental determinations performed with different types of strain gauges) - figure taken from Micro Measurements (Vishay group)

3.4 Determination of stresses and strains on wind turbine blades

The indirect method, recommended by Micro Measurements Vishay, was used to determine the principal stresses and the principal directions in the ***rectangular transducer***, measurements in Figure 3.1, which involves the following steps:

- The classical relations of deformation variation around a point (for orthotropic materials), the deformations along the main directions of the composite material are obtained:

$$\varepsilon_{xx} = \frac{1}{2}\left[\varepsilon_a + \varepsilon_b + \sqrt{2(\varepsilon_a - \varepsilon_b)^2 + (\varepsilon_b - \varepsilon_c)^2}\right] \quad (1)$$

$$\varepsilon_{yy} = \frac{1}{2}\left[\varepsilon_a + \varepsilon_b - \sqrt{2(\varepsilon_a - \varepsilon_b)^2 + (\varepsilon_b - \varepsilon_c)^2}\right] \quad (2)$$

$$\gamma_{xy} = \sqrt{2(\varepsilon_a - \varepsilon_b)^2 + (\varepsilon_b - \varepsilon_c)^2} \quad (3)$$

$$\varepsilon_{11} = \varepsilon_{xx} m^2 + \varepsilon_{yy} n^2 + 2\gamma_{xy}\, mn \quad (4)$$

$$\varepsilon_{22} = \varepsilon_{xx} n^2 + \varepsilon_{yy} m^2 - 2\gamma_{xy}\, mn \quad (5)$$

$$\gamma_{12} = (-\varepsilon_{xx} + \varepsilon_{yy})mn + \gamma_{xy}\,(m^2 - n^2) \quad (6)$$

Where the deformations $\boldsymbol{\varepsilon_{xx}}, \boldsymbol{\varepsilon_{yy}}$ and $\boldsymbol{\gamma_{xy}}$ were calculated based on the measured qualities $\boldsymbol{\varepsilon_a}, \boldsymbol{\varepsilon_b}$ *and* $\boldsymbol{\varepsilon_c}$, and m and n are: m=cosΘ; n=sinΘ, where Θ is the angle between the stress direction and the main reinforcement direction.

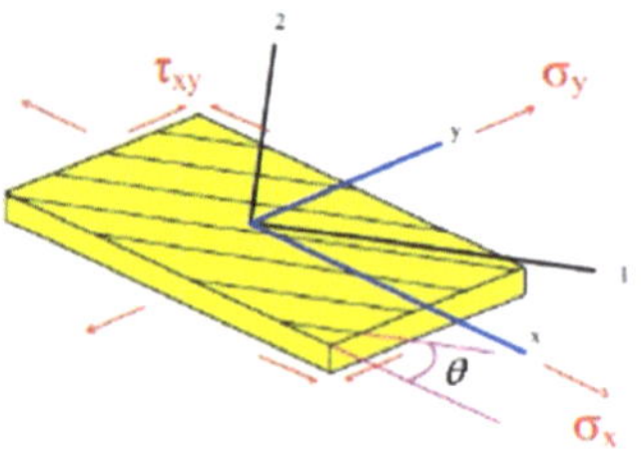

- Using stress-strain relationships, calculate the stresses in the principal axes of the material:

$$\sigma_{11} = \frac{E_{11}\,(\varepsilon_{11} + \nu_{21}\varepsilon_{22})}{1 - \nu_{12}\nu_{21}} \quad (7)$$

$$\sigma_{22} = \frac{E_{22}(\nu_{12}\,\varepsilon_{11}+\varepsilon_{22})}{1-\nu_{12}\nu_{21}} \tag{8}$$

$$\tau_{12} = G_{12}\gamma_{12} \tag{9}$$

where E_{11} = 21337 MPa, E_{22}= 18711 MPa, ν_{12}= 0.10, ν_{21}= 0.09 and G_{12}= 7824 MPa, ***were determined and presented in chapter 2 of the book TESTING OF COMPOSITE MATERIALS USED IN WIND TURBINE BLADES.***

- Calculate the principal stresses $\sigma_p = max$, $\sigma_q = min$ and their directions, the angle α

$$\sigma_p = \frac{\sigma_{11}+\sigma_{22}}{2} + \sqrt{\left(\frac{\sigma_{11}-\sigma_{22}}{2}\right)^2 + \tau_{12}{}^2} \tag{10}$$

$$\sigma_q = \frac{\sigma_{11}+\sigma_{22}}{2} - \sqrt{\left(\frac{\sigma_{11}-\sigma_{22}}{2}\right)^2 + \tau_{12}{}^2} \tag{11}$$

$$\alpha = \frac{1}{2} arctg^{-1} \frac{2\tau_{12}}{\sigma_{11}-\sigma_{22}} \tag{12}$$

The angle of the maximum principal strain direction, θ,

$$\theta = \frac{1}{2} arctg \left[\frac{2\varepsilon_b-\varepsilon_a-\varepsilon_c}{\varepsilon_a-\varepsilon_c}\right] \tag{13}$$

To determine the angle of the direction of the maximum main normal stress, θ, the following are achieved:

- The value of the angle determined by the Formula (13) is in radians and must be converted to degrees;
- Depending on the sign of numerator and denominator of the fraction in relation (13), the final angle of the maximum main deformation direction is calculated with one of the following relations:
  - If numerator ≥ 0 and denominator > 0 then

    $$\varphi = |\theta|$$

  - If numerator > 0 and denominator ≤ 0 then

    $$\varphi = 90 - |\theta|$$

  - If numerator ≤ 0 and denominator < 0 then

    $$\varphi = 90 + |\theta|$$

  - If numerator < 0 and denominator ≥ 0 then

    $$\varphi = 180 - |\theta|$$

For the ***delta rosette*** in Figure 3.5, the deformations $\boldsymbol{\varepsilon_{xx}}, \boldsymbol{\varepsilon_{yy}}$ *and* $\boldsymbol{\gamma_{xy}}$ will be calculated based on the measured $\varepsilon_a, \varepsilon_b$ *and* ε_c , but with different formulas than those presented above:

$$\varepsilon_{xx} = \frac{\varepsilon_a+\varepsilon_b+\varepsilon_c}{3} + \sqrt{\left(\frac{2\varepsilon_a-\varepsilon_b-\varepsilon_c}{2}\right)^2 + \frac{1}{3}(\varepsilon_c - \varepsilon_b)^2} \quad (14)$$

$$\varepsilon_{yy} = \frac{\varepsilon_a+\varepsilon_b+\varepsilon_c}{3} - \sqrt{\left(\frac{2\varepsilon_a-\varepsilon_b-\varepsilon_c}{2}\right)^2 + \frac{1}{3}(\varepsilon_c - \varepsilon_b)^2} \quad (15)$$

$$\gamma_{xy} = \frac{2}{\sqrt{3}}(\varepsilon_b - \varepsilon_c) \quad (16)$$

Calculations continue as above: determine ε_{11}, ε_{22} şi γ_{12}, then σ_{11}, σ_{22} σ τ_{12}, then σ_p, σ_q and Θ.

3.5 Calibration with weights under bending and torsional stress

Before the simple and compound stress tests were performed with the automation system, a calibration with weights had been performed.

The purpose of the calibrations was to determine the force-displacement and torque-rotation characteristics based on the use of calibrated weights. On the other hand, when introducing such weights, from which the force at the free end (which produces the bending moment) and the torque are derived, we followed two values: the maximum value of the displacement at the blade tip (arrow) for the bending load and the maximum value of the rotation for the twisting load.

The values obtained from the calibration were reference parameters: for the X axis the parameter was the angle that will be applied to the blade tip and for the Z axis the parameter was the displacement that will also be applied to the blade tip. These parameters will have been introduced into the EstlCam program, for the automated performance of both simple and compound loads.

The calibration with weights for bending is shown in Figure 3.13. Thus, 1 is the linear displacement transducer and 2 is the plate on which the 9 weights were successively placed. In Figure 3.14, it is observed that the tip of the linear transducer is positioned in the direction where the displacement produced by the stepper motor will be applied.

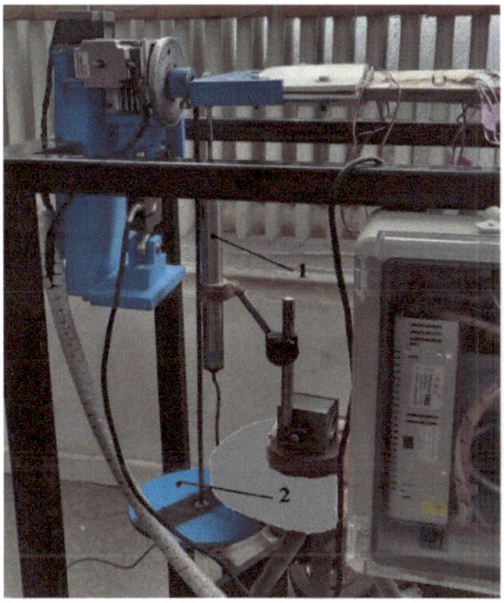

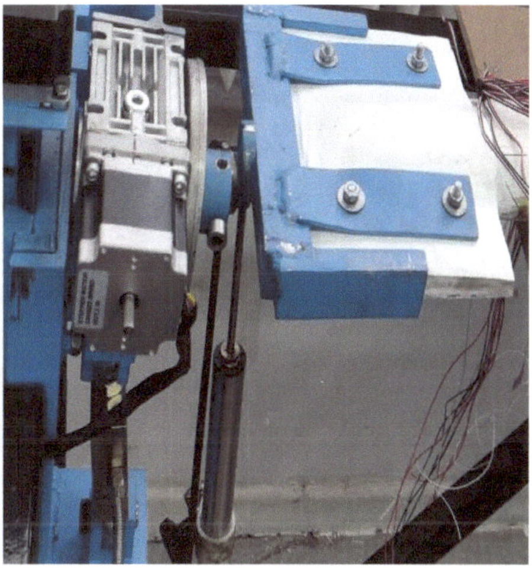

Fig. 3.13 Linear displacement sensor location

Fig. 3.14 Positioning the linear displacement sensor

In Figure 3.15, the force-displacement calibration line is plotted and its equation is: y= 5.71·displacement. The adopted maximum displacement value for the bending stress was 15 mm, for a total weight of 9.072 kg, as shown in Table 3.1. This maximum value of 15 mm was multiplied by 5.71, resulting in a maximum force of approximately 86 N. After this calibration, the maximum displacement value of 15 mm for the bending stress will be entered into the EstlCam program for all stress types.

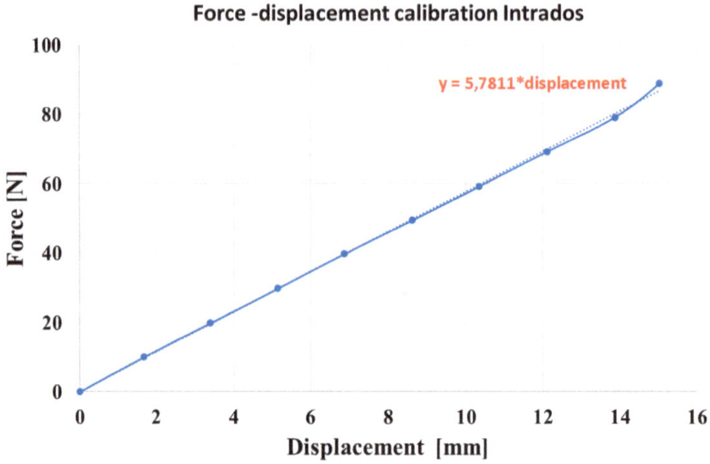

Fig. 3.15 Force-displacement calibration right

Table 3.1 presents the recorded displacement and force values at different load steps.

The displacement signal, on the other hand, is received from transducer 1, Figure 3.13, whose transformation coefficient mV/displacement is 2.44.

Table 3.1 Bending calibration table

Weights		Displacement	Force
[kg]	Mv	[mm]	[N]
0	0	0	0
1.01	41	1.6769	9.9081
1.015	83	3.3947	19.8653
1.01	126	5.1534	29.7734
1.015	168	6.8712	39.7305
0.995	211	8.6299	49.4915
1	253	10.3477	59.3015
1,015	296	12.1064	69.2586
1	339	13.8651	79.0686
1	367	15.0103	88.8786

The stiffness obtained experimentally in the bending calibration was k=F/f=5.719N/mm. The FEA calibration simulation obtained 86 N/15 mm=

5.7(3) N/mm. The relative error between the stiffness obtained experimentally and by FEA simulation is about 0.35 %.

The calibration with twist weights is shown in Figure 3.16. Thus, 1 represented the level which was placed on the arm balanced by the weights noted with 2. With the aid of the digital plotter 4, on which a plumb line was placed, it was possible to read each angle for the 7 weights that were successively placed on the plate 4. In order for the loading arm to remain horizontal, an arm was required in the opposite direction to balance the weight of the active rod and turntable.

Figure 3.17 illustrates the calibration line that has the relationship: y= 14.057·angle. The maximum adopted angle value for the twisting stress was 4° for a total weight of 7.042 kg, as shown in Table 3.2. This angle value of 4° was multiplied by 14.057, which resulted in a maximum torsional moment of approximately 56 N·m.

Aftcr this calibration, thc maximum value of the 4° angle for the twist solicitation will be entered into the EstlCam program for all solicitation types.

Fig. 3.16 Calibration with twisting weights

Figure 3.17 presents the graph for twist calibration.

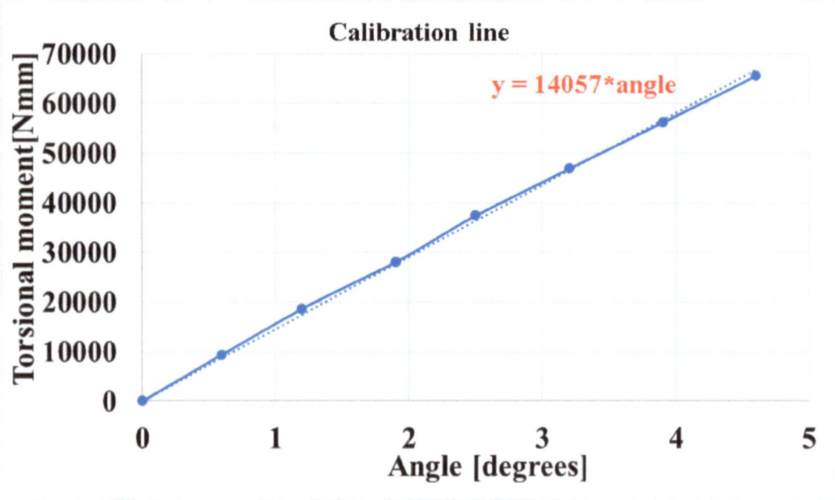

Fig. 3.17 Calibration right at twist

Table 3.2 presents the recorded values of angle versus loading moment.

Table 3.2 Angle moment values

Weights	Arm	Moment	Angle
[kg]	[mm]	[Nmm]	[degrees]
0	0	0	0
+1.005	950	9366.098	0.6
+2	950	18639	1.2
+3.01	950	28051.7	1.9
+4.015	950	37417.79	2.5
+5.03	950	46877.09	3.2
+6.03	950	56196.59	3.9
7.042	950	65627.92	4.6

3.6 Experimental results obtained on rectangular transducers

Data acquisitions from the four C2A-06-062WW-350 type transducers were performed with the Vishay P3 bridge at which the sampling rate was 1 second.

***Note**: All data acquisitions from the four transducers mentioned above, placed on the blade, as presented in the previous subsection, were performed simultaneously, using four Vishay P3 bridge, using 12 channels (3 strain gauges x 4 transducers) out of the 16 available (each bridge has 4 channels). Three tests were performed under the same conditions, with the following figures showing the results obtained from only one test. It was found that each of the three similar tests provided data that led to very similar results.*

The static bending load was applied using a stepper motor. This motor moves the tip of the blade, inducing a force perpendicular to the blade axis, which causes it to bend. Based on the calibration described above, the forces were determined step by step as a function of the blade tip displacements. The bending moment, on the horizontal axis in the graphs below, was determined as follows:

$$M_i = F \cdot l_{tr}$$

where F is the force determined previously in relation to the displacement of the blade tip and l_{tr} is the distance from the loading point (blade tip) to the transducer under consideration. For transducers 1 and 2, this distance was 1620 mm, and for transducers 3 and 4, the distance was 230 mm. Data acquisitions were performed in both directions, i.e., both during loading and unloading.

***Note:** in the following graphs, an oscillating variation will be observed for the points close together in the graph. It should be noted that the stepper motor was loaded using a screw-drill mechanism. Thus, each time the contact switches from one position to another, there is a decrease in the torque transmitted, resulting in a temporary discontinuity in the movement of the blade tip.*

The maximum displacement value for the bending load was 15 mm, which, in accordance with the previously obtained calibration coefficient, resulted in a maximum force of approx. 86 N. The maximum bending moment value at the end of the stroke was approx. 142 N·m. The maximum rotation value was 4 degrees, which led to a maximum torsional moment of approx. 56 N·m, calculated based on the calibration previously performed using weights.

Using the calculation relationships presented above, successive values were determined for both loading and unloading for:

- Maximum normal stress;
- Minimum normal stress;
- Maximum shear stress.

Below (first five figures) are the results obtained in the loading and unloading scenarios of the blade under bending stress, clockwise torsion stress, anticlockwise torsion stress, compound bending and clockwise torsion, and compound bending and anticlockwise torsion, for **transducer 1**. Thus, Figure 3.18 shows the variation of the maximum and minimum normal stresses and the maximum shear stress for transducer 1 (located at a distance of 1620 mm from the point of application of the force), both for loading and unloading under bending, in relation to the variation of the bending moment (see the calculation method in the subparagraph presenting the calibration).

From Figure 3.18, for the **bending load** and stresses determined using transducer 1, the following can be observed:

- Oscillations caused by discontinuous engagement within the screw-nut mechanism can be observed;
- During charging and discharging, the stresses increase and decrease almost continuously, which suggests correct data acquisition, except for the oscillations mentioned above. On the other hand, there is a return to zero on discharge of the values provided by the Vishay P3 bridge, which shows that there is no residual effect in terms of introducing tensions into the composite blade, but also that the automated return to zero with the stepper motor was performed correctly;
- The highest stress values, in absolute terms, are recorded for minimum stresses, with the highest value being approximately 7 MPa;
- The highest value of the maximum shear stress was approx. 2 MPa.

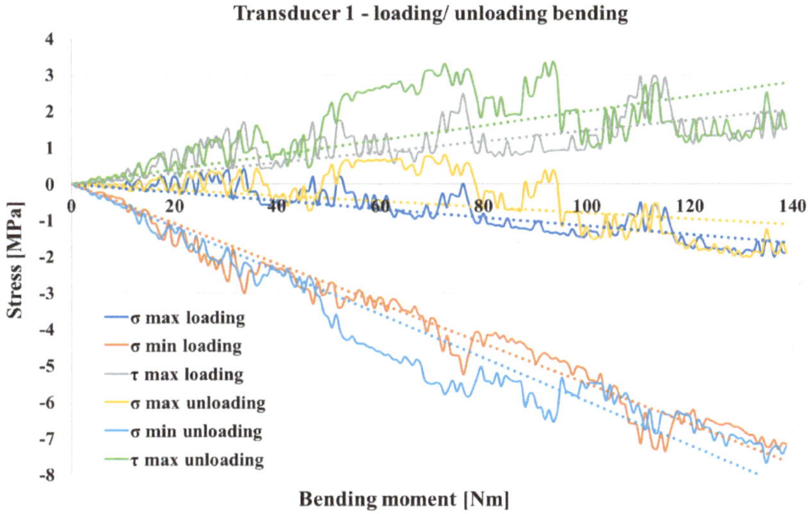

Fig. 3.18 Variation of maximum and minimum normal stresses, and maximum shear stress for transducer 1 under bending loading; loading and unloading

The torsion load was applied both clockwise and anticlockwise, relative to the blade axis viewed from the end where the load was applied. Figure 3.19 shows the variation of the maximum and minimum normal stresses and the maximum shear stress for transducer 1 (located at a distance of 1620 mm from the point of force application), both for loading and unloading under torsion, in relation to the variation of the torsional moment determined by the calibration presented above.

From Figure 3.19, for the **clockwise torsion load**, considering the stresses calculated based on the specific deformations acquired from the strain gauges corresponding to transducer 1, the following can be observed:

- The same oscillations are noticed, the causes being previously explained;
- During charging and discharging, the stresses increase and decrease almost continuously, with a return to zero on discharge of the values provided by the Vishay P3 bridge;
- The highest stress values, in absolute terms, are recorded for minimum stresses, with the highest value being approximately 6 MPa and the highest value for maximum normal stress being approximately 2 MPa;
- The highest value of the maximum shear stress was approx. 3 MPa;

- The maximum normal stress values obtained under torsional loading at the point of measurement of specific deformations with transducer 1 are lower than those obtained for the same transducer under bending loading.

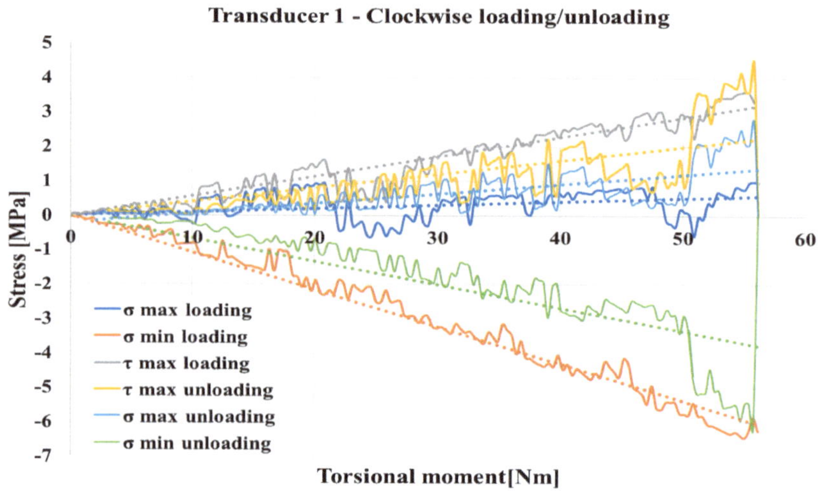

Fig. 3.19 Variation of maximum and minimum normal stresses, and maximum shear stress for transducer 1 under clockwise torsional loading; loading and unloading

Figure 3.20 presents the variation of maximum and minimum normal stresses and maximum shear stress for transducer 1, both for loading and unloading under torsion, in relation to the variation of the torsional moment.

From Figure 3.20, for the **anticlockwise torsional loading**, the following can be observed:

- The same oscillations and return to zero after discharge are observed;
- The highest values of stresses, in absolute terms, are now recorded for maximum normal stresses and are approximately 6 MPa, compared to 3 MPa in the clockwise direction;
- Also in absolute terms, the minimum stresses have lower values than those obtained for clockwise torsion, 3 MPa compared to 6.5 MPa;
- The highest value of the maximum shear stress was approx. 3 MPa, slightly lower than in the clockwise direction;

From these observations, considering that the measurements were performed on the same transducer, under the same conditions, we find that the maximum normal stress, minimum normal stress, and maximum tangential stress differ in relation to the direction of the applied torsional moment. This can also happen under operating conditions, where high stresses introduced as a result of external loads can cause the actual blade to approach its strength limit, given that an equivalent stress is determined based on the stresses from the bending load and the torsional load.

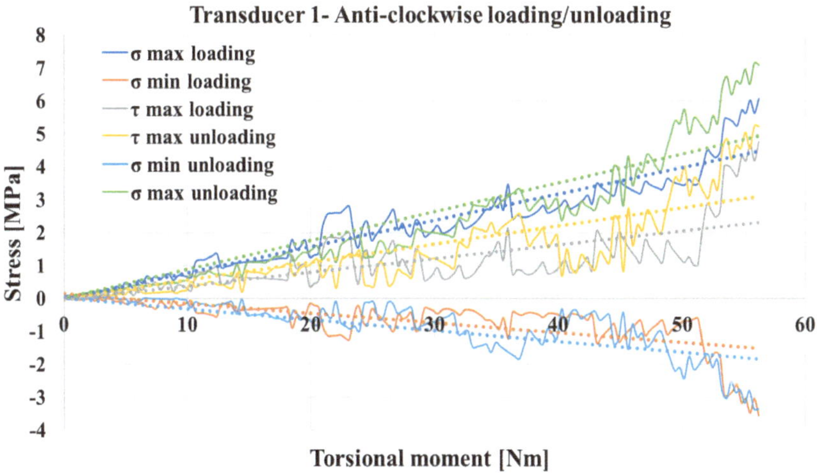

Fig. 3.20 Variation of maximum and minimum normal stresses, and maximum shear stress for transducer 1 under anticlockwise torsional loading; loading and unloading

In fact, these comparisons are presented in the following graphs, which show the compound bending and torsion stress in both variants: bending from the intrados to the extrados, to which the torsion stress is added simultaneously in the two variants, clockwise and anticlockwise.

Considering the in-situ stresses of the wind turbine blade, a combined bending and torsion test was performed on the test bench by simultaneously activating both stepper motors. The torsion loads were applied both clockwise and anti-clockwise.

Figure 3.21 shows the variation of the maximum and minimum normal stresses and the maximum tangential stress for transducer 1, both for loading and unloading under bending + torsion, in relation to the variation of a calculated compound moment, point by point, with the relationship:

$$M_c = \sqrt{M_i^2 + M_t^2}$$

where:

- M_c is the compound moment appearing on the abscissa of the following graphs;
- M_i is the bending moment calculated as shown in the previous paragraph;
- M_t is the torsional moment calculated previously based on the calibration procedure.

From Figure 3.21, for the **bending + torsion load in a clockwise direction,** considering the stresses calculated based on the specific deformations acquired from the strain gauges corresponding to transducer 1, the following can be observed:

- The highest stress values are recorded for the minimum normal stresses (in absolute value). Their maximum value is approx. 12 MPa, compared to 7 MPa for the bending load and 6 MPa for the clockwise torsion load;
- The maximum principal stress values are close to zero, which suggests the introduction of a uniaxial stress state into the blade material at the point corresponding to transducer 1, with the axis lying in the direction of the maximum principal normal stress;
- The maximum shear stress value is approximately 4 MPa.

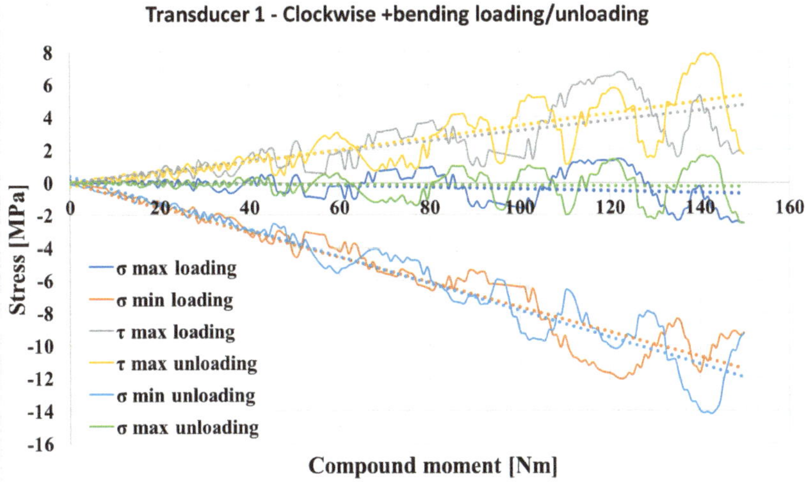

Fig. 3.21 Variation of maximum and minimum normal stresses, and maximum shear stress for transducer 1 under compound bending and clockwise torsion; loading and unloading.

From Figure 3.22, for the **bending + torsion load in the anticlockwise direction**, considering the stresses calculated based on the specific deformations acquired from the strain gauges corresponding to transducer 1, the following can be observed:

- The highest stress values are recorded for the minimum normal stresses (in absolute value). Their maximum value is approx. 10 MPa, compared to 7 MPa for the bending load and 6.5 MPa for the anti-clockwise torsion load, and compared to the highest value of the maximum principal normal stress in the case of the compound bending + torsion load in a **clockwise direction**, which is 12 MPa;
- The maximum principal stress values are close to zero but tend to increase towards the area with higher stresses, where the maximum stress reaches an approximate value of 2 MPa;
- The maximum shear stress value is approximately 5 MPa.

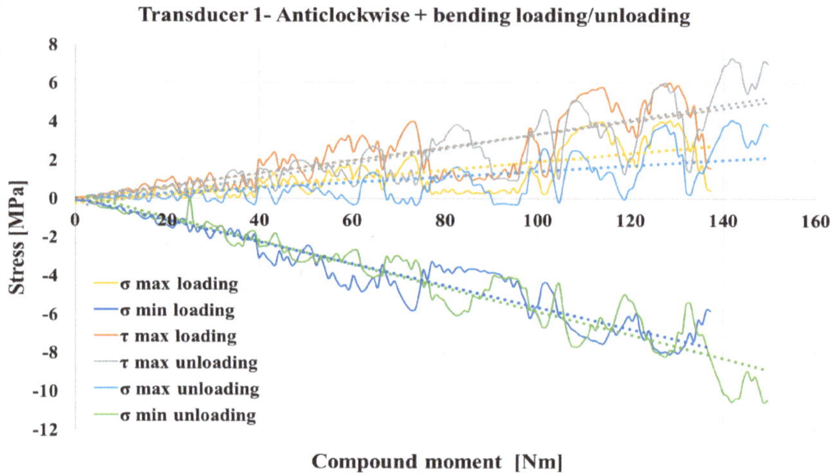

Fig. 3.22 Variation of maximum and minimum normal stresses, and maximum shear stress for transducer 1 under the compound bending + torsion load in the anti-clockwise direction; loading and unloading

Transducer 2 was placed at a distance of 1620 mm from the tip of the blade on the scale, which is also the loading point for: bending load, torsional load, both clockwise and anticlockwise, as well as compound bending and torsional loads, with torsion in both directions, in this case as well. Unlike transducer 1, transducer 2 was placed near the so-called neutral plane of the cross section, but it was placed at the bottom, where bending loads lead to compression loads. Data acquisition was performed on Vishay P3 bridges, with a sampling interval of 1 second, and data acquisition was performed both during loading and unloading.

These graphs also show data fluctuations, which, as mentioned above, are caused by the load applied by the stepper motors and the passage through the gear wheels. If, upon bending, the translational movement was achieved by the screw-nut mechanism driven by the stepper motor, upon torsion, the rotational movement was achieved by passing through two gear wheels.

Using the calculation relationships presented above, successive values were determined for both loading and unloading for:

- Maximum normal stress;

- Minimum normal stress;
- Maximum shear stress.

The following five figures show the results obtained in the loading and unloading scenarios of the blade under bending stress, clockwise torsion stress, anticlockwise torsion stress, compound bending and clockwise torsion, and compound bending and anticlockwise torsion.

Figure 3.23 presents the variation of the maximum and minimum normal stresses and the maximum tangential stress for transducer 2 (located at a distance of 1620 mm from the point of force application), both for loading and unloading under bending, in relation to the variation of the bending moment.

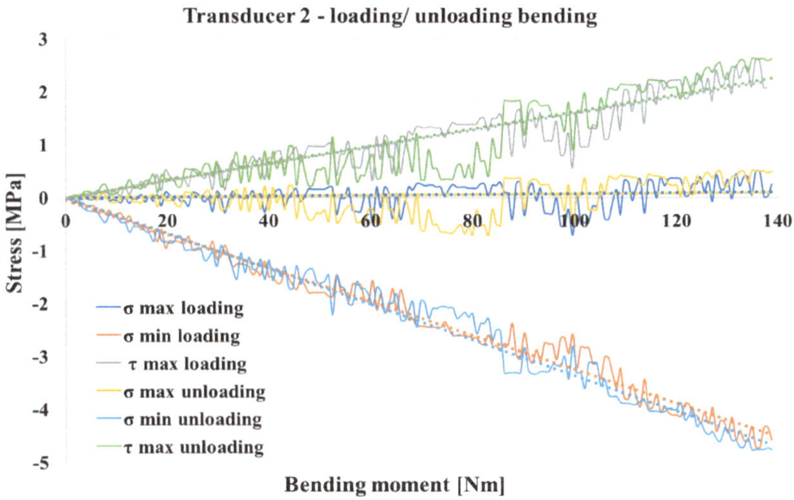

Fig. 3.23 Variation of maximum and minimum normal stresses, and maximum shear stress for transducer 2 under bending loading; loading and unloading

From Figure 3.23, for the **bending load** and the stresses determined using **transducer 2**, the following can be observed:

- The highest stress values, (at M_i=142 N·m) in absolute value, are recorded for the minimum stresses, their highest value being approximately 5 MPa, and the maximum stress values are close to zero, which leads to the conclusion that we have a state of uniaxial stress, with the axis lying in the direction of the maximum principal normal stress;

- It is found that the maximum (minimum) stress value for transducer 2 is lower than that for transducer 1 (**tr1 = 7 MPa; tr2 = 5 MPa**);
- The highest value of the maximum shear stress was approx. 2 MPa;

It is found that the values increase and decrease linearly in relation to the loading and unloading stresses, and on unloading the values return to zero.

From Figure 3.24, for the **clockwise torsional load** and the stresses determined using **transducer 2**, the following can be observed:

- The highest stress values, in absolute terms, are recorded for the minimum normal stresses, the highest value being approximately 3MPa;
- The highest values (at M_t=57 N·m) of the maximum stress are approximately 1 MPa. It can be seen that the highest value of the minimum stress for transducer 2 is lower than that corresponding to transducer 1 (**tr1 = 6 MPa; tr2=3MPa**);
- The highest value of the maximum shear stress was approximately 2 MPa.

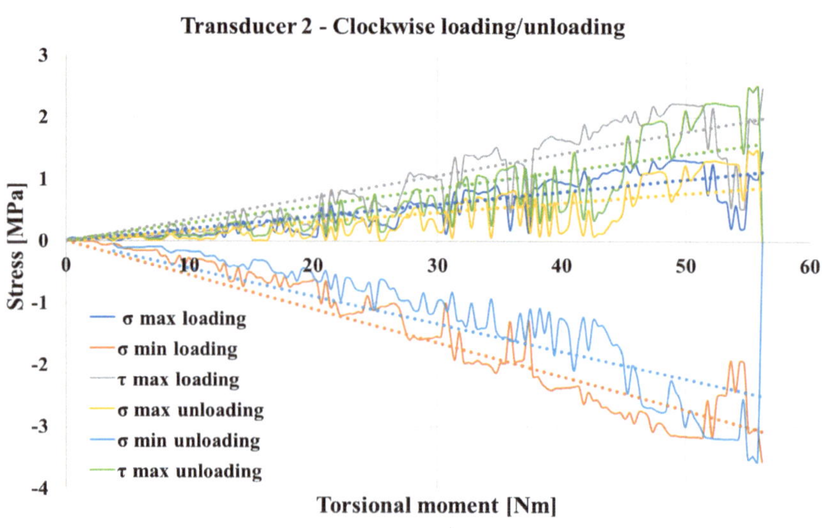

Fig. 3.24 Variation of maximum and minimum normal stresses, and maximum shear stress for transducer 2 under clockwise torsional loading; loading and unloading

From Figure 3.25, for the **anticlockwise torsional load**, based on the data provided by transducer 2 and the calculated stresses, as shown above, the following can be observed:

- The highest stress values are recorded for the maximum normal stresses and are approximately 3 MPa. For clockwise rotation, the highest stress values are recorded for the maximum and are also approximately 3 MPa;
- In absolute terms, the minimum stresses have lower values than those obtained for the clockwise torsional load, 1 MPa compared to 3 MPa;
- The highest value of the maximum shear stress was approximately 1.5 MPa, slightly lower than for the clockwise rotation (2 MPa).

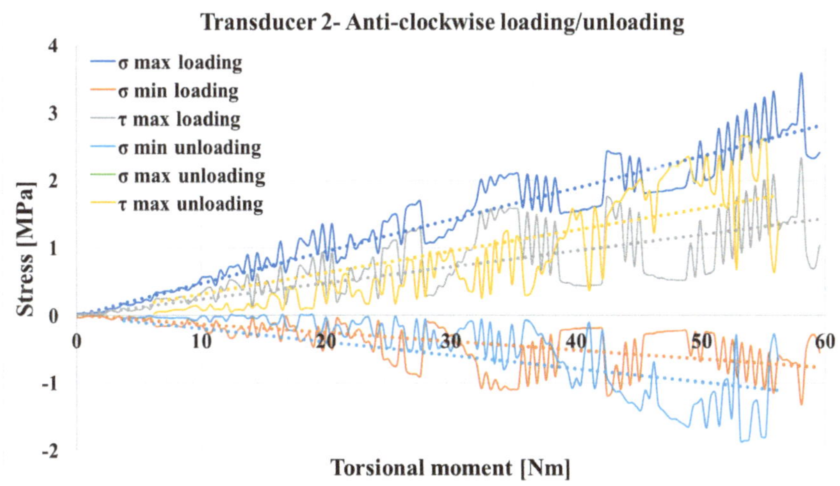

Fig. 3.25 Variation of maximum and minimum normal stresses, and maximum shear stress for transducer 2 under anticlockwise torsional loading; loading and unloading

Figure 3.26 shows the variation of the maximum and minimum normal stresses and the maximum shear stress for the **compound bending and clockwise torsion** stress on the blade, the stresses being calculated using the specific deformations taken from transducer 2, located at a distance of 1620 mm from the blade tip (the area where the bending moment is applied), at a short distance from the so-called neutral plane of the section. The following is observed:

- The minimum normal stress has a value of approx. 6 MPa at the maximum blade rotation of 4 degrees (M_t=57 Nm);

- The highest values (at M_t=57 N·m) of the maximum stress are around 0 MPa. This shows that the state of principal normal stresses introduced by the torsional load leads to a state of uniaxial stress, with the axis lying in the direction of the maximum principal normal stress;
- It can be seen that the highest value of the minimum stress for transducer 1 is lower than that for transducer 2 (**tr1 = -12 MPa; tr2=-6MPa**);
- The highest value of the maximum shear stress was approximately 2.5 MPa.

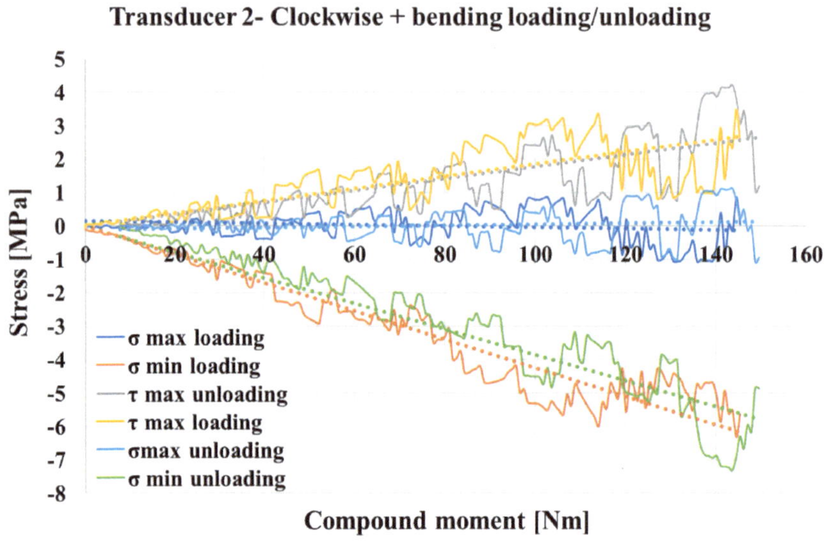

Fig. 3.26 Variation of maximum and minimum normal stresses, and maximum shear stress for transducer 2 under compound bending and clockwise torsion; loading and unloading

From Figure 3.27, for the **bending + torsional load in the anticlockwise direction**, considering the stresses calculated based on the acquisition of specific deformations from the strain gauges corresponding to transducer 2, the following can be observed:

- The highest values, at maximum blade rotation of four degrees, are recorded for minimum normal stresses (in absolute terms). Their maximum value is approximately 7 MPa, compared to the highest value of the maximum principal normal stress in the case of compound bending + torsion in a clockwise direction, which is 6 MPa. As a result, for transducer

2, there are no significant differences in the highest values of minimum stress for different blade rotation directions and, consequently, for different directions of torsional moment;

- The maximum principal stress values are close to zero, which implies that the compound bending and torsion load introduces a uniaxial principal normal stress state, with the axis lying in the direction of the maximum principal normal stress.
- The maximum shear stress value is approximately 3 MPa.

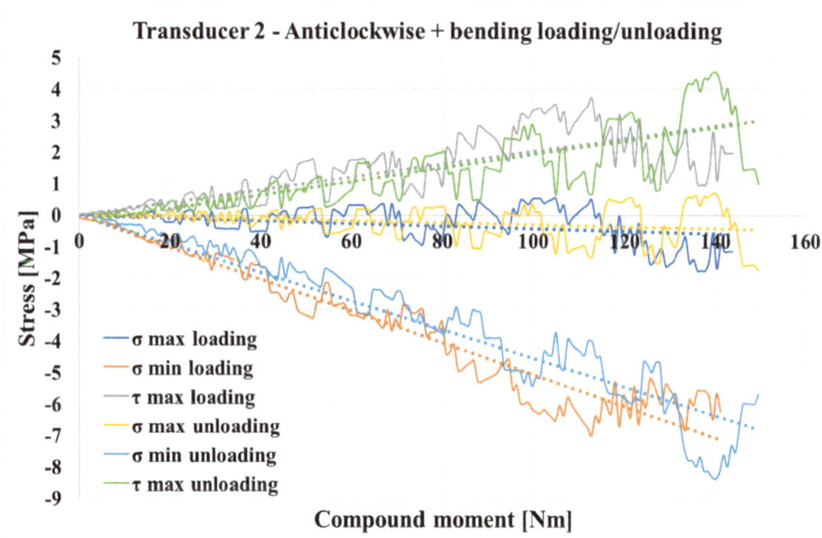

Fig. 3.27 Variation of maximum and minimum normal stresses, and maximum shear stress for transducer 2 under the compound bending + torsion load in the anti-clockwise direction; loading and unloading

Transducer 3 is located at a distance of 230 mm from the tip of the blade, respectively from the area where the loads were applied:

- bending by moving downwards (in the intrados-extrados direction) using a stepper motor via a screw-nut mechanism;
- torsion by rotating the other stepper motor in both directions using two gear reducers;
- a compound bending with torsion in both directions of torsion by combining the two.

Transducer 3 was placed slightly below the so-called neutral plane of the cross section. Data acquisition was performed on the Vishay P3 bridge at a sampling interval of 1 second, with data acquisition performed during both loading and unloading.

The data obtained by calculations in Excel using the relationships presented above, both for loading and unloading, led to the following stresses:

- Maximum normal stress;
- Minimum normal stress;
- Maximum shear stress.

The following five figures show the results obtained in the loading and unloading scenarios of the blade under bending stress, clockwise torsion stress, anticlockwise torsion stress, compound bending and clockwise torsion, and compound bending and anticlockwise torsion.

Figure 3.28 presents the variation of the maximum and minimum normal stresses and the maximum tangential stress for transducer 3 (located at a distance of 230 mm from the point of force application), both for loading and unloading under bending, in relation to the variation of the bending moment. The bending moment at the location of transducer three was: M_i=86·230=19.780 N·m.

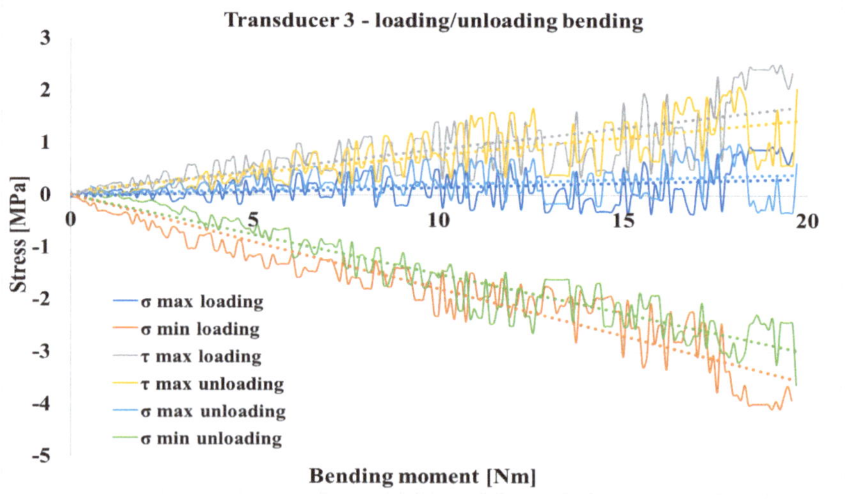

Fig. 3.28 Variation of maximum and minimum normal stresses, and maximum shear stress for transducer 3 under bending loading; loading and unloading

From Figure 3.28, for the **bending load** and stresses determined using **transducer 3**, the following can be observed:

- The highest value (in absolute terms) of the stresses was obtained for the minimum stress and is approximately 4 MPa. Here, the following remark must be made: at transducer 2, for which the bending moment is 7 times greater, the same stress (also having the highest value) was obtained as 5 MPa. The explanation for this difference lies in the fact that the section at transducer 3 (and 4 – see below) was smaller, the blade being constructed, approximately, as a **solid of equal strength;**
- The maximum stress values are close to zero, with a slight tendency to increase under load to a value of approximately 0.3 MPa;
- The shear stress value at maximum load (15 mm displacement) is approx. 1.5 MPa.

From Figure 3.29, for the **clockwise torsion** request, considering the stresses calculated based on the acquisition of specific deformations from the strain gauges corresponding to **transducer 3**, the following can be observed:

- During loading and unloading, oscillations in the obtained values are observed, these being of lower frequency than the previous ones;
- The highest stress values, in absolute terms, are recorded for minimum stresses, with the highest value being approximately 17 MPa. It should be noted that the stresses on this transducer caused by torsion are higher than on transducers 1 and 2, which are subjected to the same stress: **tr. 3 (min)=17 MPa**; **tr. 2 (min)=3 MPa**; **tr. 1(min)=6MPa**. These differences result from the different sizes and shapes of the sections in the two locations, with the maximum torsional moment being the same (approx. 57 MPa);

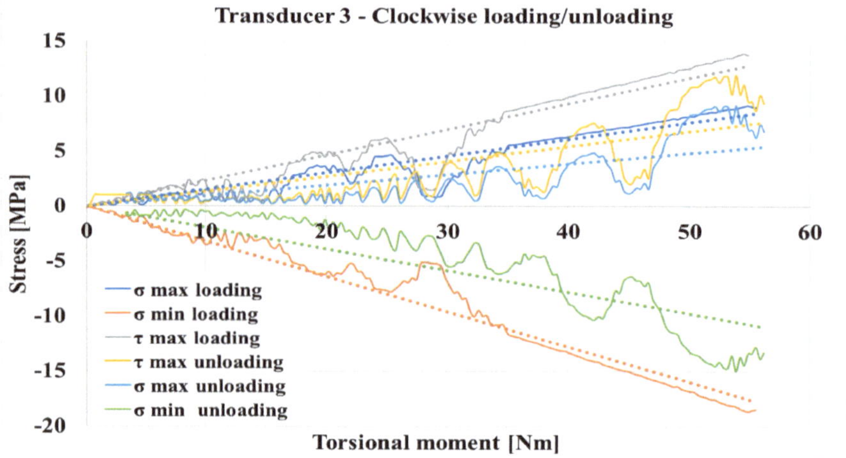

Fig. 3.29 Variation of maximum and minimum normal stresses, and maximum shear stress for transducer 3 under clockwise torsional loading; loading and unloading

- The highest value of maximum normal stresses is 8 MPa;
- The highest value of maximum shear stress was approx. 13 MPa;
- The maximum normal stress values obtained under torsional stress at the point of measurement of specific deformations with transducer 3 are much higher than those obtained for the same transducer under bending stress: torsion=8 MPa; bending=0.3 MPa.

Figure 3.30 shows the variation of the maximum and minimum normal stresses and the maximum shear stress for transducer 3, both for loading and unloading under torsion, in relation to the variation of the torsional moment, from zero to approximately 58 Nm.

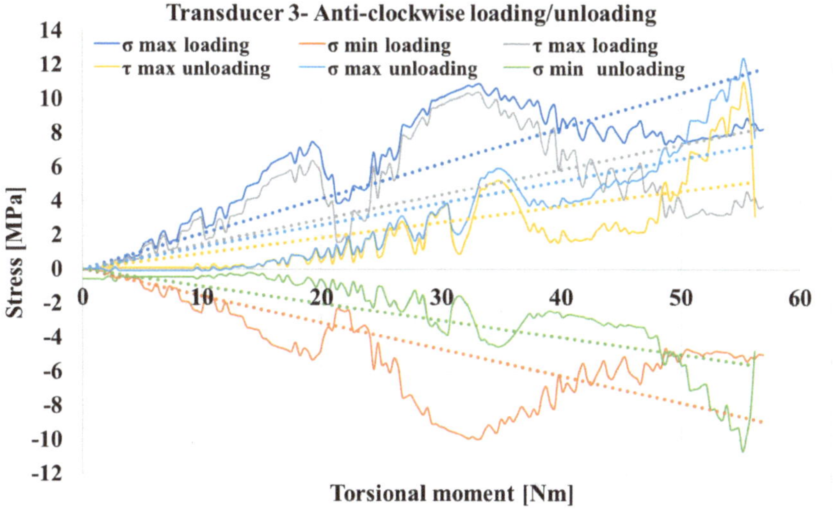

Fig. 3.30 Variation of maximum and minimum normal stresses, and maximum shear stress for transducer 3 under anticlockwise torsional loading; loading and unloading

Figure 3.30 shows the variation of the maximum and minimum normal stresses and the maximum shear stress for transducer 3, both for **anticlockwise torsional loading**. The following observations can be made:

- Both during unloading and loading, greater fluctuations in the values obtained are observed, with these fluctuations following approximately the same trend during loading and unloading;
- The highest stress values are recorded for minimum normal stresses and are approximately 9 MPa, compared to 17 MPa in the clockwise direction. Therefore, there are significant differences in the values obtained for the same load and the same transducer, the difference being between the directions of application of the torsional moment;
- The highest value of the maximum normal stress is approximately 8 MPa;
- The highest value of the maximum shear stress is approximately 9 MPa.

Considering these findings in conjunction with those from the clockwise direction, it can be seen that the normal maximum, normal minimum, and shear maximum stresses differ in relation to the direction of the applied torsional moment. This can also happen under operating conditions, where high stresses, introduced as a result of external loads, can cause the actual blade to approach its strength limit, under conditions where an equivalent stress is determined based on the stresses from the bending load and the torsional load.

From Figure 3.31, for the **bending + torsion request in a clockwise direction**, considering the stresses calculated based on the acquisition of specific deformations from the strain gauges corresponding to **transducer 3**, the following can be observed:

- The highest stress values are recorded for the minimum normal stresses (in absolute value). Their maximum value is approximately 20 MPa, compared to 4 MPa for the bending stress and 17 MPa for the clockwise torsion stress;
- The maximum principal stress values are increasing, with the maximum stress being approximately 5 MPa;
- The maximum shear stress value is approximately 9 MPa.

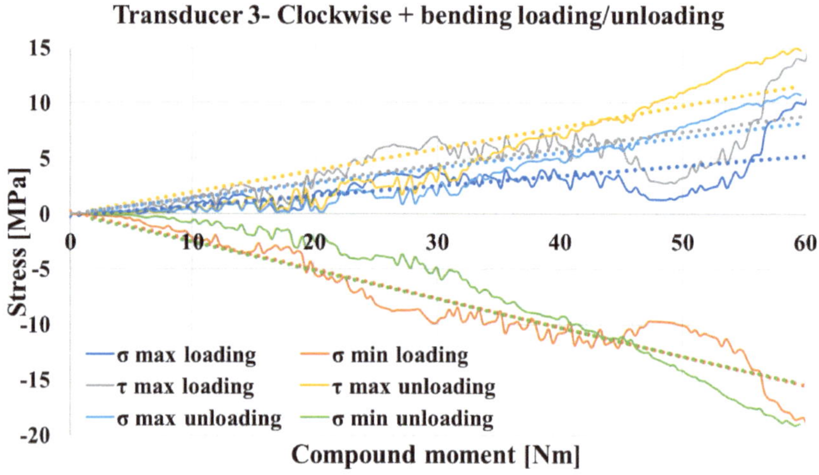

Fig. 3.31 Variation of maximum and minimum normal stresses, and maximum shear stress for transducer 3 under compound bending and clockwise torsion; loading and unloading

From Figure 3.32, for the **bending + torsion stress in the anticlockwise direction**, considering the stresses calculated based on the acquisition of specific deformations from the strain gauges corresponding to **transducer 3**, the following can be observed:

- The highest stress values are recorded for maximum normal stresses. Their maximum value is approximately 15 MPa, compared to values of 0.3 MPa for bending load and 8 MPa for anticlockwise torsion load, and compared to the highest value of minimum principal normal stress in the case of combined bending + clockwise torsion load, which is 20 MPa;

- The maximum principal stress values are relatively high compared to the previous ones for the same stress, being approximately 15 MPa;
- The same can be said about the maximum values of the maximum shear stress, which are approximately 11 MPa.

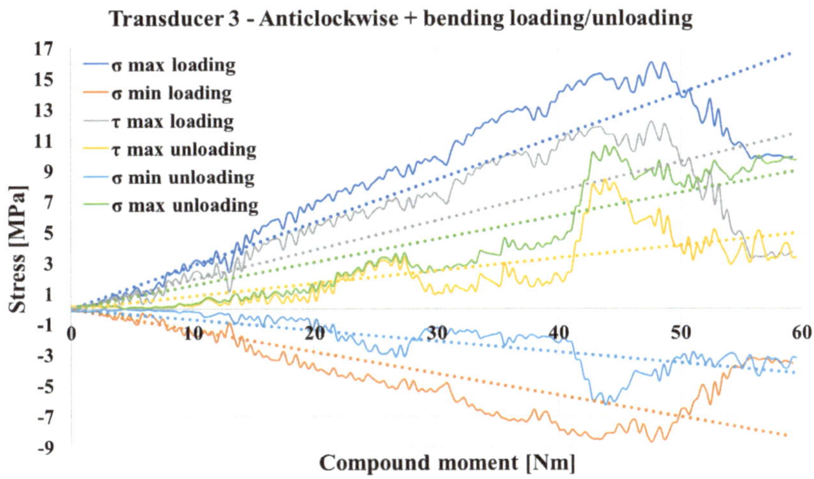

Fig. 3.32. Variation of maximum and minimum normal stresses, and maximum shear stress for transducer 3 under the compound bending + torsion load in the anticlockwise direction; loading and unloading

Transducer 4 was placed at the same distance from the end of the blade (loading point) as transducer 3. Transducer 4 was placed at the bottom of the cross section, where we have compressive stresses due to bending.

Figure 3.33 shows the variation of the maximum and minimum normal stresses and the maximum shear stress for transducer 4 (located 230 mm from the point of force application) for both bending loading and unloading, in relation to the variation of the bending moment.

From Figure 3.33, for the **bending load** (15 mm displacement of the free end of the blade) and the stresses determined using **transducer 4**, the following can be observed:

- The highest value (in absolute terms) of the stresses was obtained for the minimum stress and is approximately 3 MPa. At transducer 3, the same value was 4 MPa. Note that the distance on the vertical section between transducer 3 and transducer 4 is approximately 10 mm;
- The maximum stress values oscillate around zero, which implies a uniaxial stress state;
- The value of the shear stress at maximum load (15 mm displacement) is approximately 1.5 MPa, comparable to that obtained for transducer 3 under the same load conditions.

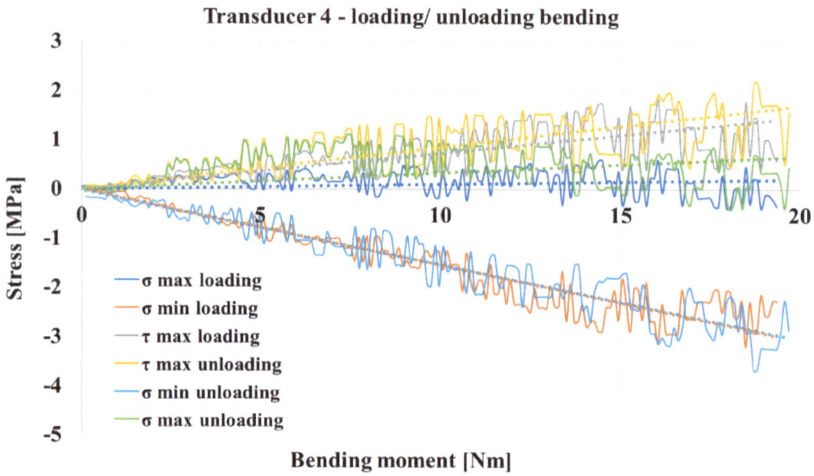

Fig. 3.33 Variation of maximum and minimum normal stresses, and maximum shear stress for transducer 4 under bending loading; loading and unloading

The results obtained for the three stresses mentioned above, for **transducer 4** subjected to **clockwise torsion**, are shown in Figure 3.34. We note the following:

- At the end of the load test, there were some unexplained drops in stress. This was probably due to issues with the torque transmission. For this reason, we will consider the representative values for this test to be the highest ones around 42 Nm for the moment of torsion;
- The highest stress values are recorded for maximum stresses, with the highest value being approximately 27 MPa and the highest value of minimum normal stress (in absolute terms) being approximately 10 MPa. It should be noted that the stresses on this transducer caused by torsion are

higher than on transducers 1, 2, and 3, which are subjected to the same stress: **tr. 4 (max)=27 MPa; tr. 3 (min)=17 MPa; tr. 2 (min)=3 MPa; tr. 1(min)=6MPa**. The differences between transducers 1 and 2 on the one hand and 3 and 4 on the other hand are due to the different sizes and shapes of the sections at the two locations. Between transducers 3 and 4, located in the same cross section, the differences stem from the different locations, per section, where the two transducers were placed;

- The highest value of the maximum shear stress was approx. 18 MPa.

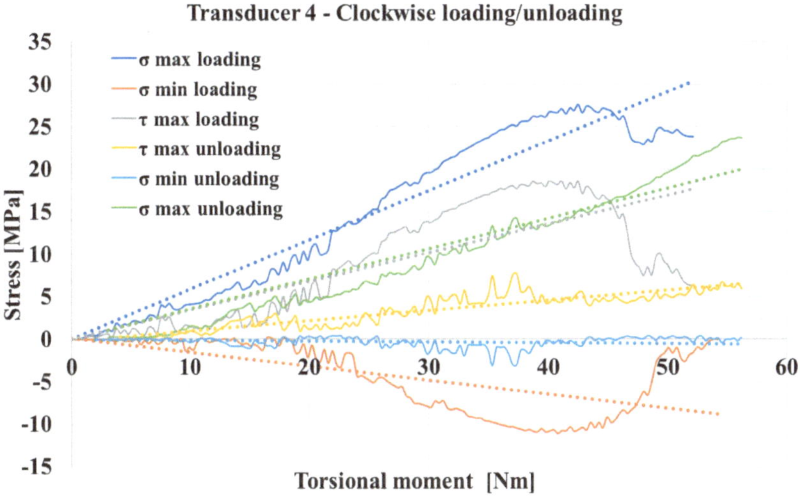

Fig. 3.34 Variation of maximum and minimum normal stresses, and maximum shear stress for transducer 4 under clockwise torsional loading; loading and unloading

Figure 3.35 shows the graphs of the variation of the aforementioned stresses in the case of torsional stress, in an **anticlockwise** direction, calculated for the specific deformations provided by transducer 4. The torsional moment is approximately 58 Nm.

The following conclusions can be drawn from observing the variations below:

- Both during unloading and loading, larger oscillations of the obtained values are observed, some oscillations following approximately the same trend during loading and unloading, others being in counter-phase;

- The highest stress values, in absolute terms, are recorded for minimum normal stresses and are approximately 40 MPa, compared to 27 MPa in the clockwise direction (maximum normal stress). There are, therefore, significant differences in the values obtained for the same load and the same transducer, but in different directions. For the anticlockwise direction at transducer 3, the highest stress value (maximum normal stress) was 8 MPa. Therefore, for the same direction of rotation, at the two transducers, 3 and 4, located in the same cross section, there are very different values, the difference being due to the location of the two transducers;
- The highest value of the maximum shear stress is approximately 30 MPa, much higher than that obtained for the same transducer but for clockwise rotation (18 MPa).

Considering these findings in conjunction with those from the clockwise direction, on the same transducer, under the same conditions, we find that the maximum normal, minimum normal, and maximum shear stresses differ in relation to the direction of the applied torque.

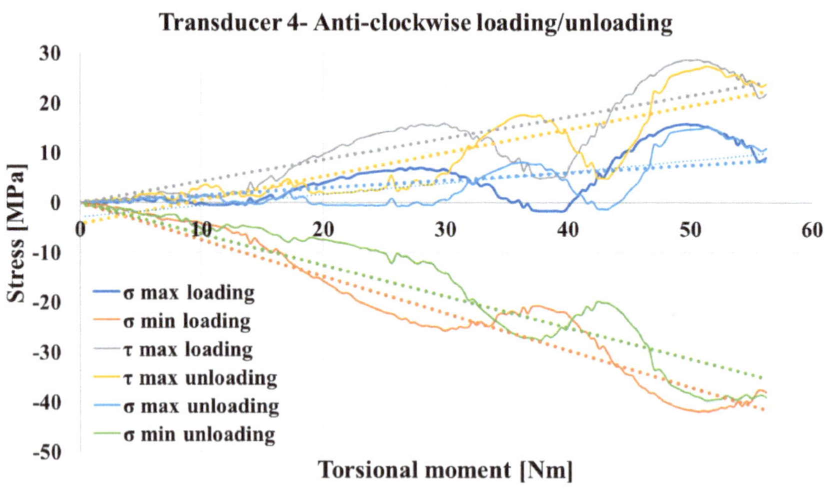

Fig. 3.35 Variation of maximum and minimum normal stresses, and maximum shear stress for transducer 4 under anticlockwise torsional loading; loading and unloading

From Figure 3.36, for the **bending + torsion request in a clockwise** direction, considering the stresses calculated based on the acquisition of specific deformations from the strain gauges corresponding to transducer 4, the following is observed:

- The same oscillations of the values obtained are recorded, resulting from the inconsistency of the load due to the play in the torque transmission mechanism: gear engagement and wedge clamping of the shaft;
- The highest stress values are recorded for maximum normal stresses. Their highest values are approximately 38 MPa, compared to the value of 20 MPa from transducer 3, under the same load conditions;
- The minimum normal stress values in absolute terms are approximately 10 MPa;
- If we compare with the other two transducers, we have: **tr1(min)=12 MPa; tr2(min)=6 MPa; tr3(min)=20 MPa and tr4 (max)=38 MPa;**
- The highest value of the maximum shear stress is approximately 25 MPa.

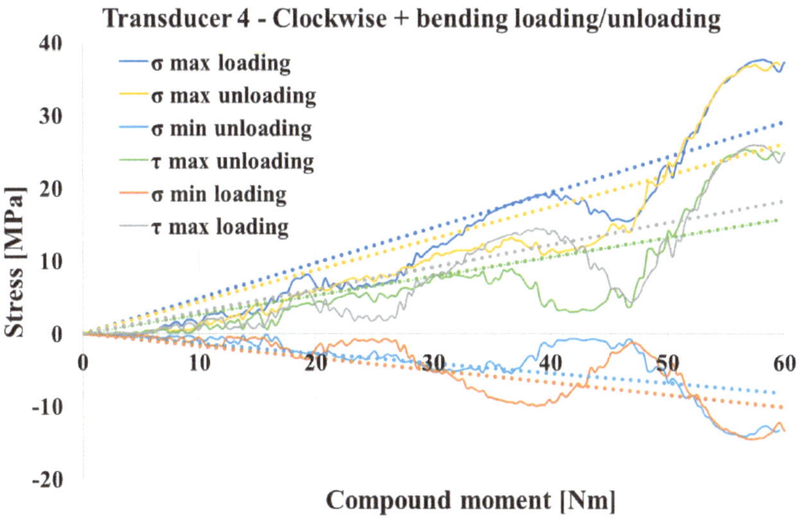

Fig. 3.36 Variation of maximum and minimum normal stresses, and maximum shear stress for transducer 4 under compound bending and clockwise torsion; loading and unloading

Figure 3.37 shows the variations in maximum normal stress, minimum normal stress, and maximum shear stress for **bending + torsion stress in the anticlockwise** direction, based on the acquisition of specific deformations from the strain gauges associated with transducer 4, from which the following can be observed:

- The highest stress values are recorded for minimum normal stresses, with an absolute value at maximum stress of approximately 50 MPa. The highest value of the maximum normal stresses is approximately 8 MPa;
- If we also consider the highest values of normal stresses for all transducers, under the conditions of combined bending + torsion (counterclockwise) stress, we will have: **tr1(min)=10 MPa; tr2(min)=7 MPa; tr3(min)=20 MPa and tr4 (min)=50 MPa;**
- For the same combined stress but with clockwise rotation, we have: **tr1(min)=12 MPa; tr2(min)=6 MPa; tr3(min)=20 MPa and tr4 (max)=38 MPa;**
- The highest value of the maximum shear stress is approximately 30 MPa (25 MPa was the result for the combined stress with clockwise rotation).

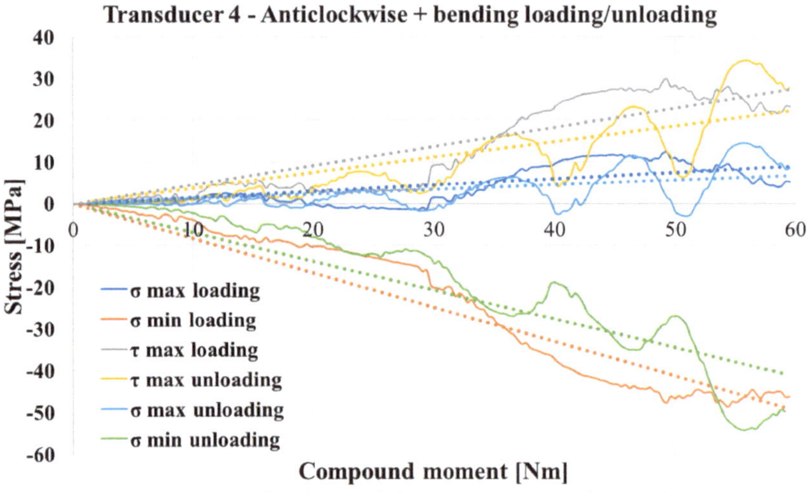

***Fig. 3.37** Variation of maximum and minimum normal stresses, and maximum shear stress for transducer 4 under the compound bending + torsion load in the anticlockwise direction; loading and unloading*

Conclusions on determining principal stresses using resistive electrical strain gauges

The tables below provide the highest values of maximum normal stress, Table 3.3, minimum normal stress, Table 3.4, and maximum shear stress, Table 3.5, for the four transducers in the case of the five loading scenarios: bending, clockwise torsional rotation, anticlockwise torsional rotation, combined bending and clockwise torsional rotation, and combined bending and anticlockwise torsional rotation. As it can be seen from the values presented above, the highest normal stress value is approximately 50 MPa, with a shear stress of 30 MPa. Expecting similar values in the compound stress and for comparison purposes, we kept the same maximum load values for the simple stresses. As a result, the simple loading produced low stress values. However, the following comparisons are useful for drawing conclusions about the stresses introduced into the composite material of the wind turbine blade at scale. Certain variations cannot be established in relation to a specific working parameter (e.g., the variation in maximum normal stress for all four transducers or the variation in maximum normal stress for one transducer but for the five load scenarios) for the following reasons:

- the bending moment differs in size (the force arm differs) for transducers 1 and 2 compared to transducers 3 and 4;
- for these pairs of transducers, both the size and the shape of the cross-section differ as well;
- significant differences were found in obtaining the results for the two directions of rotation in the case of torsional loading;
- each of the four transducers was placed in a different location; even if, for example, transducers 3 and 4 were placed in the same cross-section, one was mounted at the maximum distance from the so-called neutral plane of the section.

For these reasons, graphs based on the data in the following tables would not be relevant.

Table 3.3 Maximum normal stress values

	Load	Transducer 1	Transducer 2	Transducer 3	Transducer 4
		Maximum normal stress value [MPa]			
1	Bending	*-1.89*	*0.26*	*0.94*	*0.59*
2	Clockwise torsional	*0.99*	*1.45*	*9.11*	*23.87*
3	Anticlockwise torsional	*6.07*	*2.41*	*10.87*	*15.74*
4	Bending and clockwise torsion	1.46	0.86	10.90	36.52
5	Bending and anticlockwise torsion	4.06	-1.14	16.04	12.41

Table 3.4 Minimum normal stress values

	Load	Transducer 1	Transducer 2	Transducer 3	Transducer 4
		Minimum normal stress value [MPa]			
1	Bending	*-7.16*	*-4.56*	*-3.90*	*-2.28*
2	Clockwise torsional	*-6.43*	*-3.55*	*-18.52*	*-11.11*
3	Anticlockwise torsional	*-3.57*	*-0.55*	*-9.97*	*-41.95*
4	Bending and clockwise torsion	-12.05	-6.31	-19.12	-14.45
5	Bending and anticlockwise torsion	-5.80	-6.12	-8.72	-48.66

Table 3.5 Maximum shear stress values

	Load	Transducer 1	Transducer 2	Transducer 3	Transducer 4
		Maximum shear stress value [MPa]			
1	Bending	*1.51*	*2.37*	*2.34*	*1.85*
2	Clockwise torsional	*3.55*	*2.47*	*13.80*	*18.57*
3	Anticlockwise torsional	*4.76*	*1.04*	*10.39*	*28.72*
4	Bending and clockwise torsion	6.75	3.49	14.96	24.02
5	Bending and anticlockwise torsion	5.99	3.72	12.21	29.85

From the analysis of the values obtained, we can draw the following conclusions:

- For simple loads, the stress values are generally lower than for compound loads. There are certain differences related to the positioning of the transducers, in this sense, for transducers 3 and 4 the stresses are higher. Although the bending moment is higher for transducers 1 and 2, there is a difference in the axial modulus of resistance, W_z between the two sections, hence the small differences in stresses obtained for the 4 transducers for bending load;
- The highest value of a stress for simple load is 42 MPa and it is obtained for the anticlockwise torsional load at transducer 4;

- The highest value of a stress for compound load is 49 MPa and it is obtained for the anticlockwise torsional bending load at transducer 4;
- The torsional stresses are generally higher than the bending stresses;
- The anticlockwise torsional stresses, also included in the compound stresses, are higher than the clockwise stresses for the same transducers.

3.7. Experimental results obtained on delta-type strain gauge chains

3.7.1 Blade bending load

As previously shown, the transducer chains each contain 5 rosettes, each with three grids positioned at 0°/120°/240°, with the transducers placed 5 mm apart from each other. The acquisition of specific strains from the 4 chains (20 electrical transducers × 3 strain gauges per transducer = 60 channels) was carried out using four QUANTUM HBM MX1615B data acquisition systems, connection cables, and a laptop with Catman-AP data acquisition software. The measurements were performed in dynamic mode, with a sampling frequency of 20 Hz, and the connection in the software interface was of the type "sg 3 wire 120".

All data acquisitions from the four chains (a number of 60 strain gauges) mentioned above, placed on the blade as presented in the previous subsection, were carried out simultaneously, both during loading and unloading, in five loading scenarios: bending, clockwise torsion, anticlockwise torsion, combined bending with clockwise torsion, and combined bending with anticlockwise torsion.

Three tests were performed under the same conditions, with the following figures showing only the results from a single test. It was found that each of the three similar tests provided data that led to very close results. Details regarding the placement of the strain gauge chains have been provided earlier. For the bending load case, the chains were positioned as follows: L1, L2, and L3 on the intrados, where the stresses due to bending are tensile, and L4 on the extrados, where the stresses due to bending are compressive.

Figure 3.6 shows the arrangement of these strain gauge rosette chains: L1 (intrados) and L4 (extrados) are positioned at a distance of 400 mm, L2 is located at 560 mm, and L3 is located at 1500 mm from the free end of the blade, where the bending and torsional loads are applied. The distance between the rosettes is 5 mm, which resulted in a change in the force arm and, consequently, in the bending moment for each individual rosette. Therefore, the bending moment is different for each rosette.

The tests were carried out as follows:

- Bending by displacement (maximum of 15 mm, corresponding to a force of 86 N, obtained through the calibration mentioned earlier) in the vertical direction and in the direction from intrados to extrados, using a stepper electric motor driving a screw–nut mechanism, software-controlled, as described previously;
- Torsion by rotation in both directions (maximum rotation was 4 degrees, corresponding to a torsional moment of 56 Nm, obtained through the calibration mentioned earlier) using a software-controlled stepper electric motor and transmission through two gear reducers to increase transmission power.

Given the large volume of data, the high acquisition frequency (20 Hz), and the oscillations that occurred due to the transmission through the gear reducer, the following will be observed in the subsequent graphs, both for stresses and for the angle of the principal normal stress:

- A certain scatter of the stress results;
- Oscillation around a trend line for the angle.

In any case, what is important is the trend highlighted by the corresponding straight lines, with the loads applied exclusively within the elastic range.

Using the calculation formulas presented earlier, successive values were determined, both for loading and unloading, for:

- Maximum normal stress;
- Minimum normal stress;
- Maximum shear stress;
- Angle of the principal normal stress.

Given that the results from unloading were similar to those from loading, all the graphs below correspond to the loading phase, while for illustration purposes, only the graphs from the first rosette of each chain are presented for the unloading phase. Since the variation angles of the principal normal stress direction exhibit roughly the same pattern, only the variations for the first rosette of each chain are shown for illustration purposes. Figure 3.38 shows the variation of the three stresses mentioned earlier for the bending load case, during loading, corresponding to **rosette 1 in chain 1**. The following observations can be made:

- During loading, the stresses increase continuously, which suggests correct data acquisition, except for the scatter previously mentioned;
- The highest stress values are recorded for the maximum normal stresses, with the largest value being approximately 3.7 MPa;
- The minimum stress value is close to the maximum stress value, which suggests a state close to biaxial tensile stress;
- Since the maximum and minimum normal stresses are close in value, the shear stress value approaches zero.

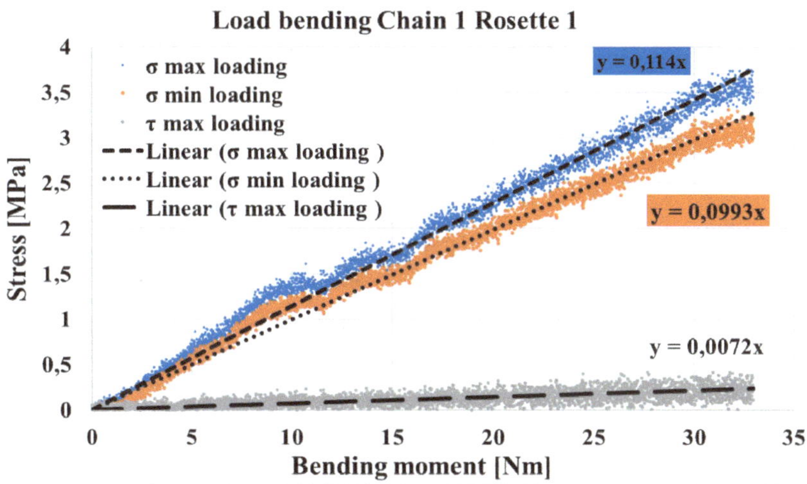

Fig. 3.38 Variation of maximum and minimum normal stresses, and maximum shear stress, for rosette 1 in chain 1 under bending load – loading phase

Figure 3.39 shows the variation of the angle of the maximum principal normal stress as the load increases from zero to its maximum value. It can be observed that the trend is for the angle to remain around zero degrees. As noted earlier, the principal stresses are both positive and of similar magnitude, which leads to the conclusion that the direction of the maximum principal normal stress is close to the direction of the gauge aligned along direction A, as shown in Figure 3.5. It should be noted that the direction of strain gauge A was along the longitudinal axis of the blade. The fluctuations that appear are caused by the loading method, which used the motion of the stepper motor and the power transmission through the gear reducer.

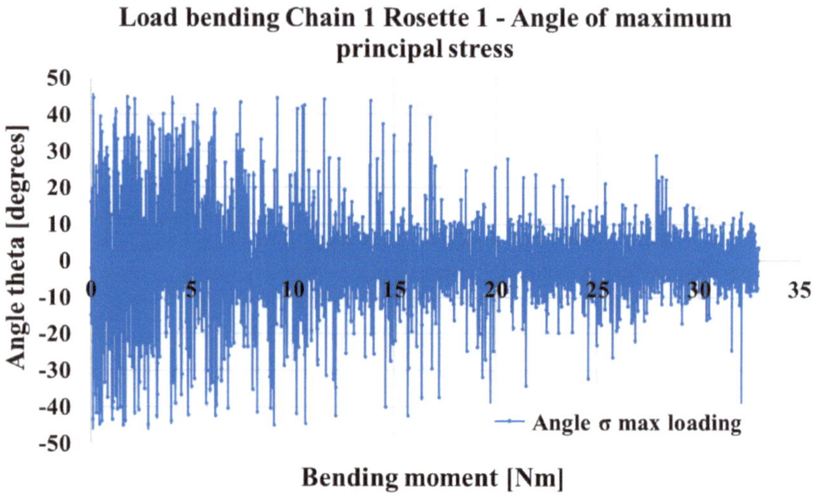

Fig. 3.39 Variation of the angle of the maximum principal normal stresses with respect to axis A, for rosette 1 in chain 1 under bending load – loading phase

Figure 3.40 shows the variation of the previously mentioned stresses during **unloading** in bending. Considering that, in all unloading cases, a similar trend was observed—passing through zero and reaching negative stress values—it can be assumed that, upon reaching zero load, the blade undergoes a slight deformation in the unloading direction, resulting in negative stresses. On the other hand, it was possible that after unloading, some residual stresses persist; that is, small positive stress values could persist once the load ceased, as it was the case throughout the test. This did not happen, which leads us to hypothesize that no residual stresses occur after loading in the case of the GFRP material from which the scaled turbine blade is made.

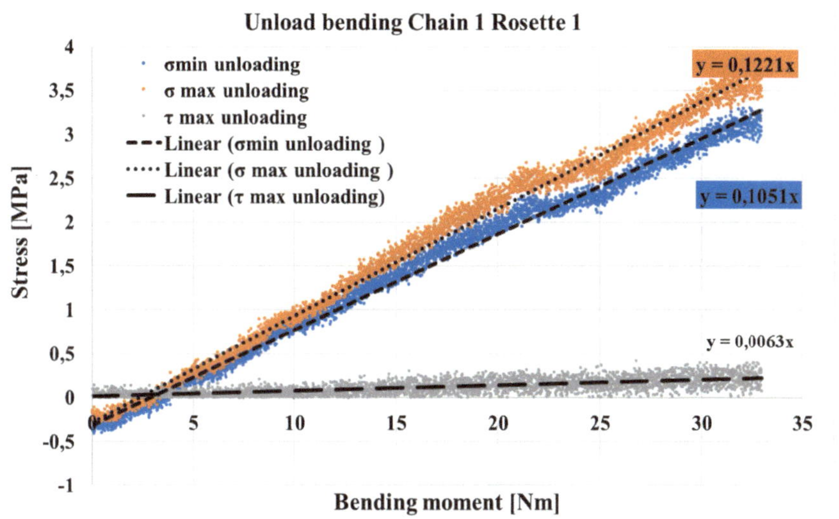

Fig.3.40 Variation of maximum and minimum normal stresses, and maximum shear stress, for rosette 1 in chain 1 under bending load – unloading phase

From Figure 3.41, which shows the variation of the angle of the maximum principal normal stress, it can be observed that its values remain around zero degrees, as it was the case during loading. During unloading, the maximum principal normal stress values also approach the minimum values.

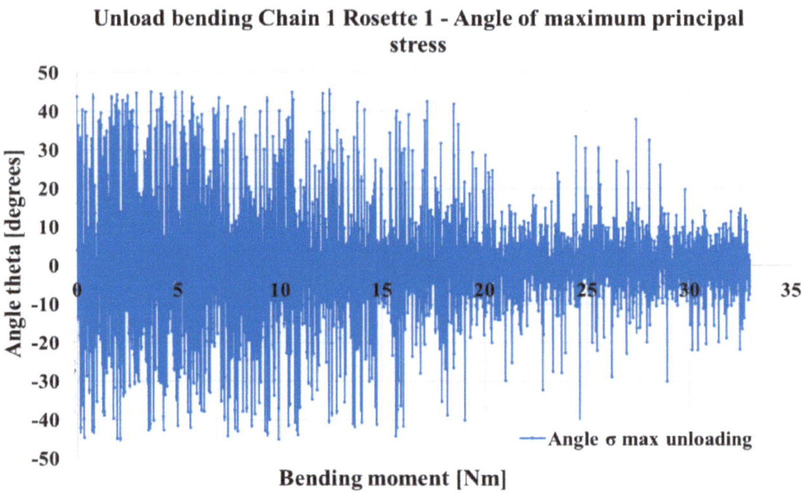

Fig. 3.41 Variation of the angle of maximum normal stresses with respect to axis 1, rosette 1 in chain 1 under bending load – unloading phase.

In Figure 3.42, the variation of the three stresses — principal normal, maximum and minimum, and maximum shear — is shown for bending loading, corresponding to strain gauge **rosette 2 from chain 1**. Rosette 2 is located at a slightly greater distance from the blade tip compared to rosette 1, the distance between the rosettes being 5 mm. The following observations can be made:

- Although there are many acquisition points, the values are arranged approximately linearly as the load increases;
- The highest stress values are recorded for the maximum stresses, the largest of these being about **4.3 MPa** (compared to **3.7 MPa** for rosette 1). It should be noted, however, that even if we move away from the blade tip — thus increasing the lever arm and, consequently, the bending moment — the stresses will not necessarily increase, because the axial moment of inertia W_z also increases as the cross-sectional dimensions of the blade grow. As it will be seen in the other stress results, the stress values do not increase linearly with distance from the point of application of the force. Considering that the cross-sectional dimensions of the blade increase toward the hub, it can be said that the blade resembles the behavior of a body of equal strength under bending;

- It is observed that the value of the minimum stress — about 3.75 MPa — is close to the value of the maximum stress, which suggests a state close to biaxial tensile stress;
- The shear stress values are close to zero.

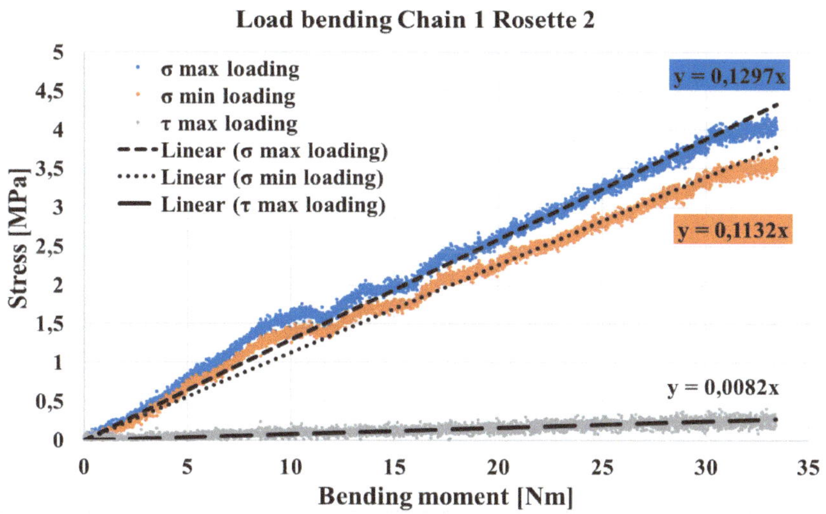

Fig. 3.42 Variation of maximum and minimum normal stresses, and maximum shear stress, for rosette 2 in chain 1 under bending load – loading phase

In Figure 3.43, the variation of the angle of the maximum principal normal stress is shown for bending loading, using data obtained from rosette 2 in chain 1. It can be observed that the values are clustered around zero degrees, just as in the case of rosette 1. Since the same behavior will occur for the other rosettes in chain 1, as well as for the rosettes in the other chains, these variations of the maximum principal normal stress angle will no longer be presented in the future.

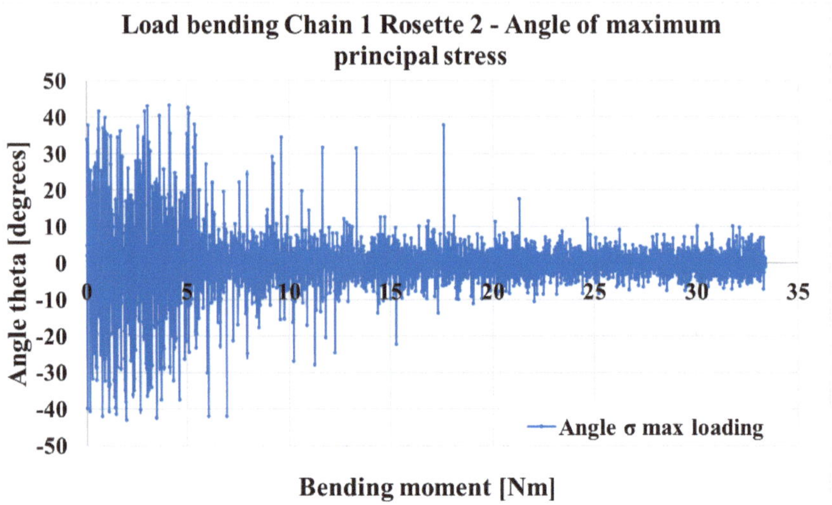

Fig. 3.43 Variation of the angle of the maximum principal normal stresses with respect to axis A, for rosette 2 in chain 1 under bending load – loading phase

In Figure 3.44, the variation of the previously mentioned stresses — principal normal, maximum and minimum, and maximum shear — is shown for **bending loading**, using data obtained from **rosette 3 in chain 1**. Rosette 3 is positioned in continuation of rosette 2, at a greater distance from the blade tip than rosette 2. The following observations can be made:

- The same small oscillations around the linear variation are observed as for rosettes 1 and 2, near the values of **12 Nm** and **16 Nm**. Since the data were collected under the same loading conditions, it is possible that these oscillations are due to nonconformities in the loading chain, namely small contact losses in the gear reducer;
- The highest stress values are recorded for the maximum normal stresses, the largest of these being about 4.2 MPa (3.7 MPa for rosette 1 and 4.3 MPa for rosette 2). It should be noted, however, that even if we move away from the blade tip — thus increasing the lever arm and, consequently, the bending moment — the stresses will not necessarily increase, because the axial moment of inertia W_z also increases as the cross-sectional dimensions of the blade grow. As it will be seen in the other stress results, the stress values do not increase linearly with the distance from the point of application of the force. Considering that the cross-sectional dimensions of

the blade increase toward the hub, it can be said that the blade resembles the behavior of a body of equal strength under bending;

- It is observed that the value of the minimum stress about 3.5 MPa is close to the value of the maximum stress, which suggests a state close to biaxial tensile stress;
- The shear stress values are close to zero.

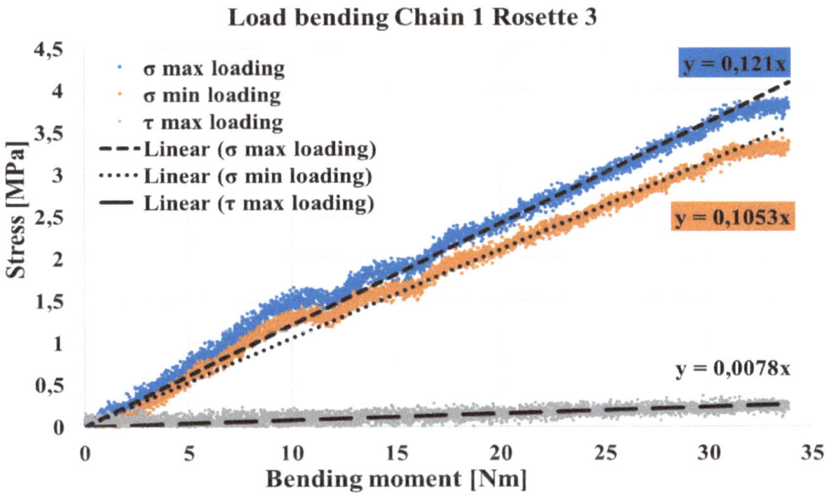

Fig. 3.44 Variation of maximum and minimum normal stresses, and maximum shear stress, for rosette 3 in chain 1 under bending load – loading phase

In Figure 3.45, the variation of the principal normal stresses — maximum and minimum — and the maximum shear stress is shown for **bending loading**, using data obtained from **rosette 4 in chain 1**. Rosette 4 is positioned in continuation of rosette 3, at a distance 5 mm greater than rosette 3. The following observations can be made:

- The same small oscillations around the linear trend as for rosettes 1 and 2 are observed, near the values of 12 Nm and 16 Nm;
- The highest stress values are recorded for the maximum normal stresses, the largest of these being about **4.8 MPa** (**3.7 MPa** for rosette 1, **4.3 MPa** for rosette 2, and **4.2 MPa** for rosette 3);
- It is observed that the value of the minimum stress about 3.3 MPa is close to the value of the maximum stress, which suggests a state close to biaxial tensile stress;

- The shear stress values are close to zero.

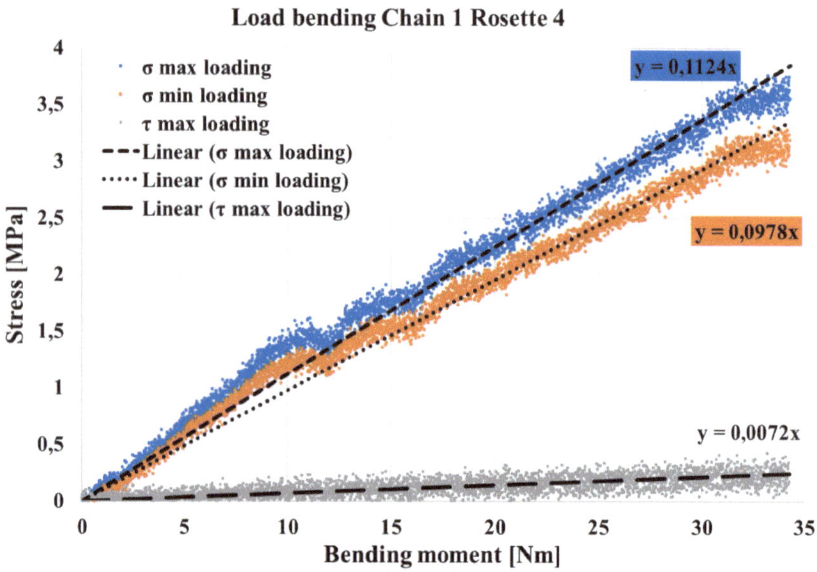

Fig. 3.45 Variation of maximum and minimum normal stresses, and maximum shear stress, for rosette 4 in chain 1 under bending load – loading phase

In Figure 3.46, the variation of the principal normal stresses, maximum and minimum, and the maximum shear stress is shown for **bending loading**, using data obtained from **rosette 5 in chain 1**. The following observations can be made:

- The highest stress values are recorded for the maximum normal stresses, the largest of these being about **4.1 MPa** (**3.7 MPa** for rosette 1, **4.3 MPa** for rosette 2, **4.2 MPa** for rosette 3, and **4.8 MPa** for rosette 4). Although the bending moment here is slightly greater than at rosette 3, for example, the stresses are comparable due to the increase in the axial section modulus W_z in the cross-section corresponding to rosette 5;
- It is observed that the value of the minimum stress — about 3.6 MPa — is close to the value of the maximum stress, which suggests a state close to biaxial tensile stress;
- The shear stress values are close to zero.

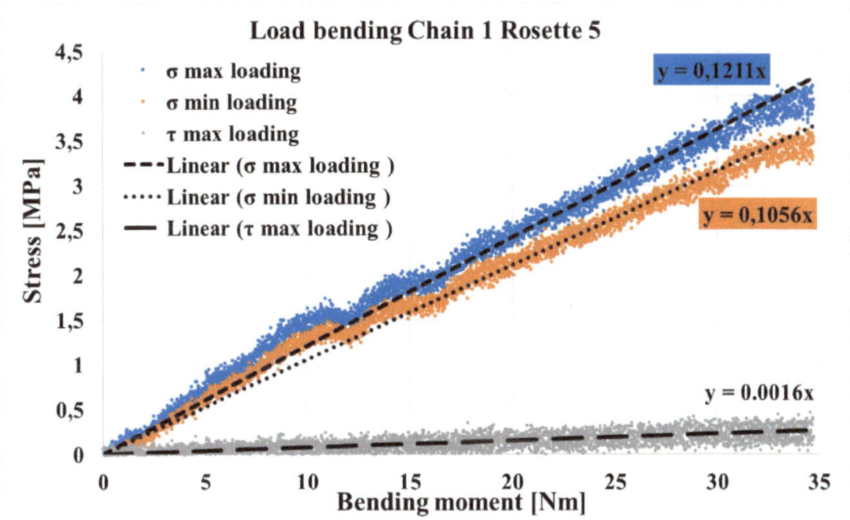

Fig. 3.46 Variation of maximum and minimum normal stresses, and maximum shear stress, for rosette 5 in chain 1 under bending load – loading phase

In Figure 3.47, the variation of the previously mentioned stresses — principal normal, maximum and minimum, and maximum shear is shown for **bending loading**, using data obtained from **rosette 1 in chain 2**. As previously mentioned, chain 2 is located 560 mm from the blade tip, where the displacement for bending is applied. The distances between rosettes within the same chain are 5 mm apart. Under these conditions, the bending moments differ both from those in chain 1 and between the rosettes within chain 2 itself. The following observations can be made:

- The highest stress values are recorded for the maximum normal stresses, the largest of these being about **3.1 MPa** (for chain 1: **3.7 MPa** for rosette 1, **4.3 MPa** for rosette 2, **4.2 MPa** for rosette 3, **4.8 MPa** for rosette 4, and **4.1 MPa** for rosette 5). The bending moment for chain 1 (with the respective variations from one rosette to another) was about 35 Nm, while for chain 2 (with the same note on variations) it is about 48 Nm. Nevertheless, as it can be seen from the results for the maximum normal stress, this value is lower for rosette 1 in chain 2 than for the rosettes in chain 1. The explanation is the one given earlier regarding the increase in W_z for the cross-section corresponding to the rosettes in chain 2;

- It is observed that the value of the minimum stress about 2.6 MPa is close to the value of the maximum stress, which suggests a state close to biaxial tensile stress;
- The shear stress values are close to zero.

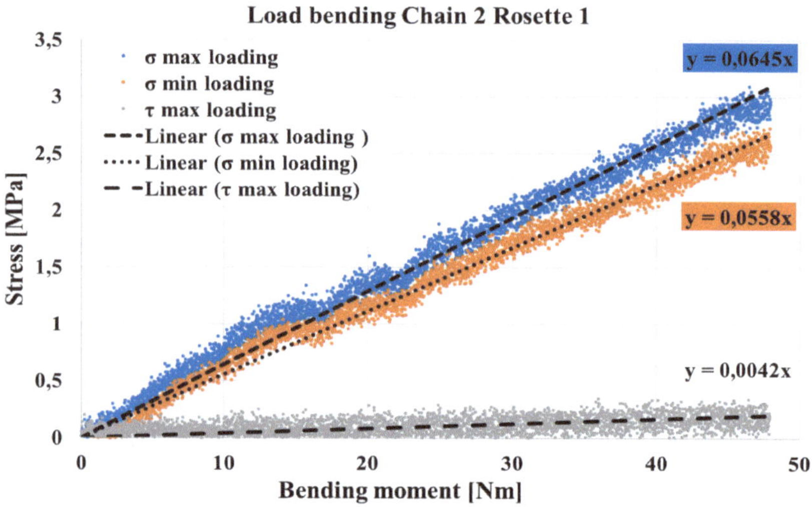

Fig. 3.47 Variation of maximum and minimum normal stresses, and maximum shear stress, for rosette 1 in chain 2 under bending load – loading phase

In Figure 3.48, the variation of the maximum principal normal stress angle is shown for bending loading, using data obtained from rosette 1 in chain 2. It is observed that the values are centered around zero degrees, as in the other cases. Since the same behavior will occur for the other rosettes in chain 2, as well as for the rosettes in the other chains, these variations of the maximum principal normal stress angle will not be presented further.

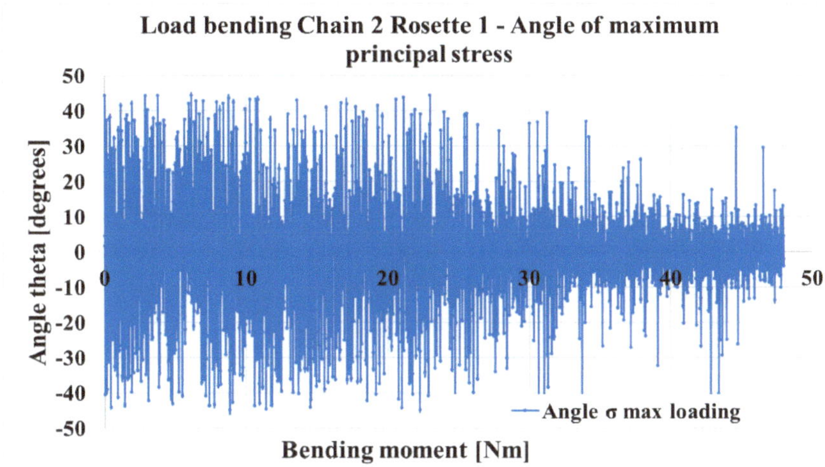

Fig. 3.48 Variation of the angle of the maximum principal normal stresses with respect to axis A, for rosette 1 in chain 2 under bending load – loading phase

In Figure 3.49, the variation of the stresses during bending **unloading** is shown. It is observed, here as well, that a complete return to zero of the specific strains (on the basis of which the stresses in the figure below were calculated) does not occur. The explanation, given earlier, is that the blade still deforms slightly in the unloading direction, resulting in negative stresses, which are not shown here.

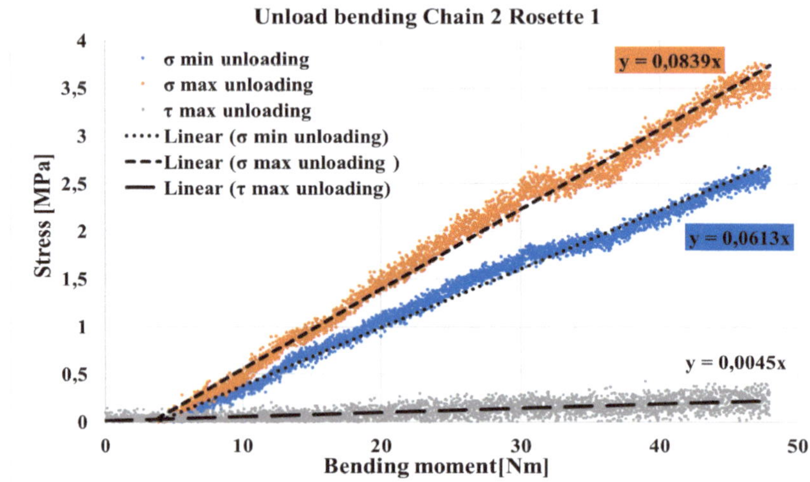

Fig. 3.49 Variation of maximum and minimum normal stresses, and maximum shear stress, for rosette 1 in chain 2 under bending load – unloading phase

Obviously, during unloading (Mi = 0), because the stresses enter the negative range, the angle of the maximum principal normal stress deviates slightly from zero, as it can be observed in Figure 3.50.

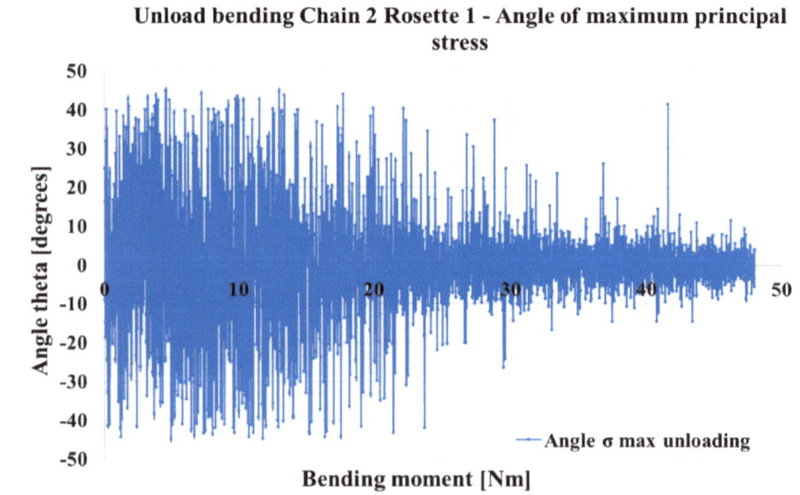

Fig. 3.50 Variation of the angle of maximum normal stresses with respect to axis 1, rosette 1 in chain 2 under bending load – unloading phase

In Figure 3.51, the variation of the three stresses principal normal, maximum and minimum, and maximum shear is shown for **bending loading**, using data obtained from **rosette 2 in chain 2**. Rosette 2 is located slightly further from the blade tip compared to rosette 1, with a distance of 5 mm between the rosettes. The following observations can be made:

- The highest stress values are recorded for the maximum stresses, the largest of these being about **3.1 MPa** (**3.1 MPa** was also recorded for rosette 1 in chain 2);
- It is observed that the value of the minimum stress, about 2.6 MPa, is close to the value of the maximum stress, which suggests a state close to biaxial tensile stress;
- The shear stress values are close to zero.

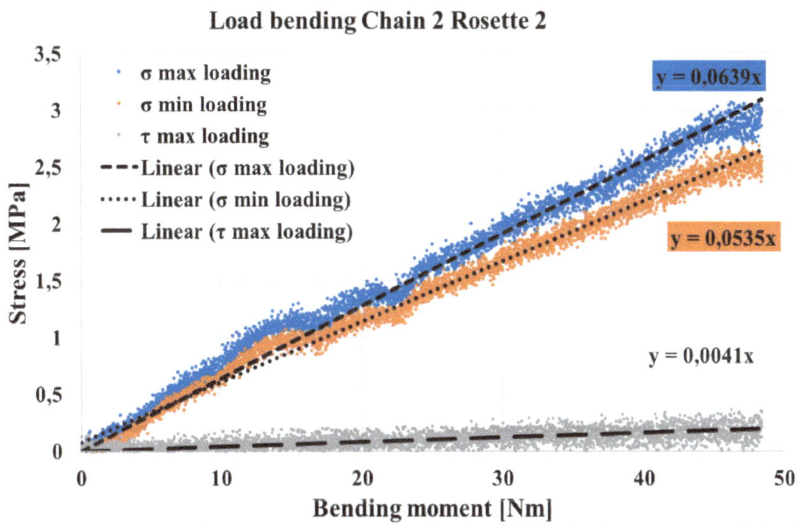

Fig. 3.51 Variation of maximum and minimum normal stresses, and maximum shear stress, for rosette 2 in chain 2 under bending load – loading phase

In Figure 3.52, the variation of the three stresses principal normal, maximum and minimum, and maximum shear is shown for **bending loading**, using data obtained from **rosette 3 in chain 2**. Rosette 3 is located slightly further from the blade tip compared to rosette 2, with a distance of 5 mm between the rosettes. The following observations can be made:

- The highest stress values are recorded for the maximum stresses, the largest of these being about **2.9 MPa** (**3.1 MPa** was recorded for rosettes 1 and 2 in chain 2);
- It is observed that the value of the minimum stress, about 2.5 MPa, is close to the value of the maximum stress, which suggests a state close to biaxial tensile stress;
- The shear stress values are close to zero.

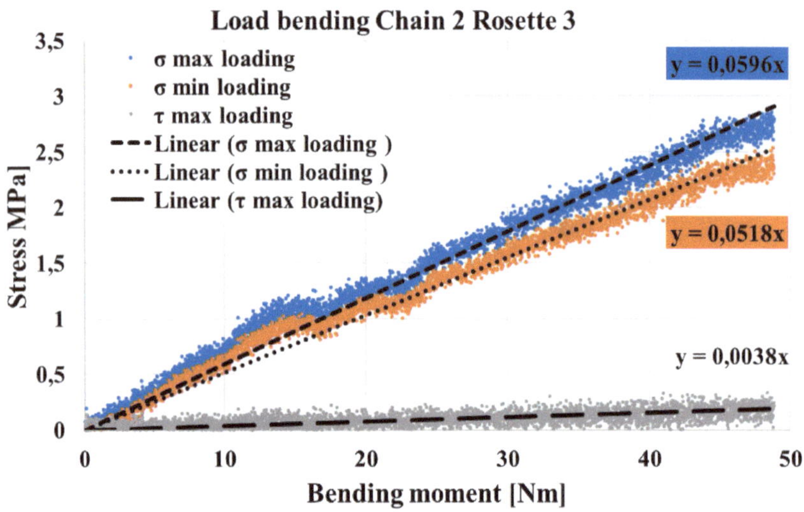

Fig. 3.52 Variation of maximum and minimum normal stresses, and maximum shear stress, for rosette 3 in chain 2 under bending load – loading phase

In Figure 3.53, the variation of the three stresses principal normal, maximum and minimum, and maximum shear is shown for **bending loading**, using data obtained from **rosette 4 in chain 2**. Rosette 4 is located slightly further from the blade tip compared to rosette 3, with a distance of 5 mm between the rosettes. The following observations can be made:

- The highest stress values are recorded for the maximum stresses, the largest of these being about **3 MPa** (**3.1 MPa** was recorded for rosettes 1 and 2 in chain 2, and **2.9 MPa** for rosette 3 in chain 2);
- It is observed that the value of the minimum stress, about **2.6 MPa**, is close to the value of the maximum stress, which suggests a state close to biaxial tensile stress;

- The shear stress values are close to zero.

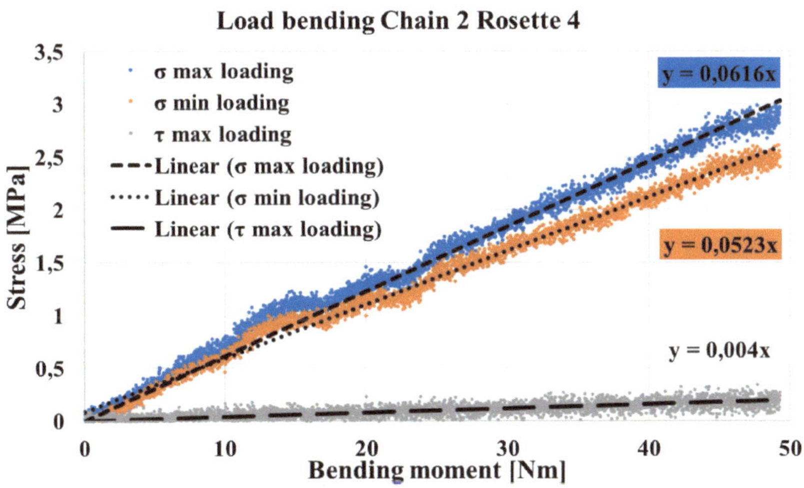

Fig. 3.53 Variation of maximum and minimum normal stresses, and maximum shear stress, for rosette 4 in chain 2 under bending load – loading phase

Figure 3.54 shows the variation of the three stresses, maximum and minimum normal stresses and maximum shear stress, under **bending load** during loading, corresponding to **rosette 5 in chain 2**. Rosette 5 is located slightly farther from the blade tip compared to rosette 4, with a distance of 5 mm between the rosettes. The following observations can be made:

- The highest stress values are recorded for the maximum normal stresses, with the largest value being approximately **3.2 MPa** (**3.1 MPa** for rosettes 1 and 2 in chain 2, **2.9 MPa** for rosette 3 in chain 2, and **3 MPa** for rosette 4 in chain 2);
- The minimum stress, approximately 2.6 MPa, is close to the maximum stress value, suggesting a state close to biaxial tensile stress;
- The shear stress values are close to zero.

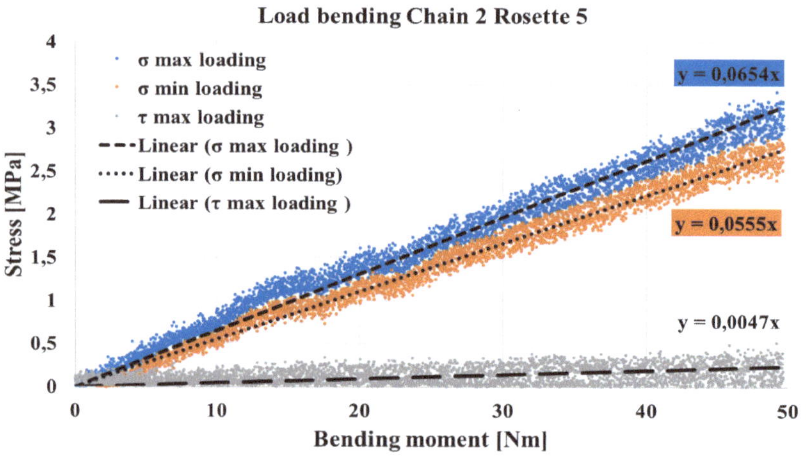

Fig. 3.54 Variation of maximum and minimum normal stresses, and maximum shear stress, for rosette 5 in chain 2 under bending load – loading phase

Figure 3.55 shows the variation of the previously mentioned stresses, maximum and minimum normal stresses and maximum shear stress, under **bending load** during loading, corresponding to **rosette 1 in chain 3**. As mentioned earlier, chain 3 is located 1500 mm from the blade tip, where the bending displacement is applied. The distances between the rosettes within the same chain are 5 mm. Under these conditions, the bending moments differ both compared to chains 1 and 2 and between the rosettes within the same chain, in this case, chain 3. The following observations can be made:

- The highest stress values are recorded for the maximum normal stresses, with the largest value being approximately **5 MPa** (for rosette 1 in chain 1, it was **3.7 MPa**, and for rosette 1 in chain 2, it was **3.1 MPa**). The bending moment for chain 1 (with the usual differences from one rosette to another) was approximately 35 Nm, for chain 2 it was approximately 48 Nm, while for chain 3 (with the same observation regarding differences), the bending moment is approximately 130 Nm. As it can be seen from the resulting maximum normal stress values, it is slightly higher for rosette 1 in chain 3 than the values obtained for the rosettes in chains 1 and 2. It is evident that the bending moment is much higher for chain 3, even though Wz is also larger in the section corresponding to the strain-gauge rosettes of chain 3;
- The minimum stress, approximately 4.2 MPa, is close to the maximum stress value, suggesting a state close to biaxial tensile stress;

- The shear stress values are close to zero.

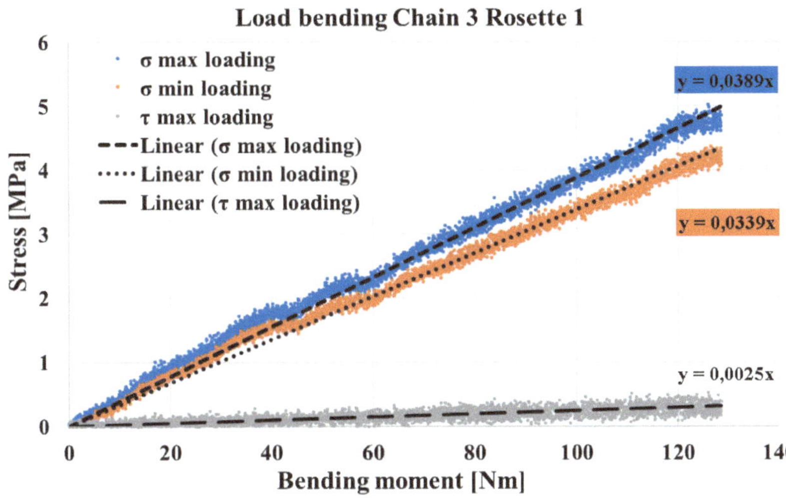

Fig. 3.55 Variation of maximum and minimum normal stresses, and maximum shear stress, for rosette 1 in chain 3 under bending load – loading phase

From Figure 3.56, whose graphs are provided for **unloading** for rosette 1 in chain 3, it can again be observed that the values of the specific strains (here, the calculated stresses) do not return perfectly to zero. However, the differences are small, for stresses less than 0.2 MPa.

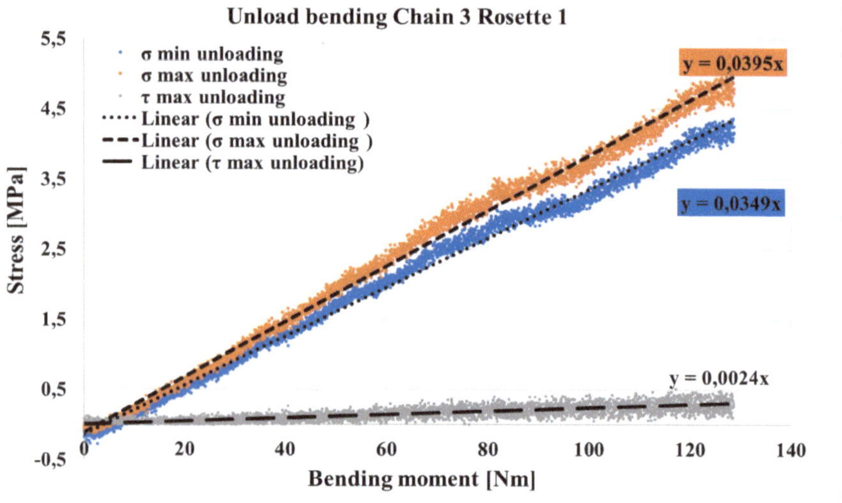

Fig. 3.56 Variation of maximum and minimum normal stresses, and maximum shear stress, for rosette 1 in chain 3 under bending load – unloading phase

In Figure 3.57, the variation of the three stresses—maximum and minimum principal normal stresses and maximum shear stress—for **bending loading**, at the applied load, corresponding to **rosette 2 in chain 3,** is presented. Rosette 2 is located slightly farther from the blade tip compared to rosette 1, with a distance of 5 mm between the rosettes. The following can be observed:

- The highest stress values occur for the maximum normal stresses, with the largest value being approximately **3.4 MPa** (**5 MPa** was recorded for rosette 1 in chain 3);
- The minimum stress, which is approximately 2.8 MPa, is close to the maximum stress, suggesting an approach to a biaxial state of tensile stresses;
- The shear stress values are close to zero.

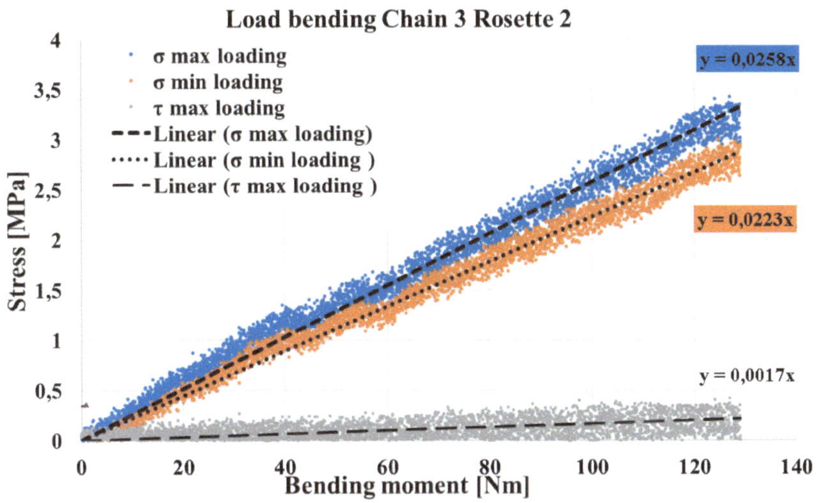

Fig. 3.57 Variation of maximum and minimum normal stresses, and maximum shear stress, for rosette 2 in chain 3 under bending load – loading phase

In Figure 3.58, the variation of the three stresses, maximum and minimum principal normal stresses and maximum shear stress, for **bending loading**, at the applied load, corresponding to **rosette 3 in chain 3,** is presented. Rosette 3 is located slightly farther from the blade tip compared to rosette 2, with a distance of 5 mm between the rosettes. The following can be observed:

- The highest stress values occur for the maximum normal stresses, with the largest value being approximately **3.6 MPa** (**5 MPa** was recorded for rosette 1 in chain 3, and **3.4 MPa** for rosette 2 in chain 3);
- The minimum stress, which is approximately 3.2 MPa, is close to the maximum stress, suggesting an approach to a biaxial state of tensile stresses;
- The shear stress values are close to zero.

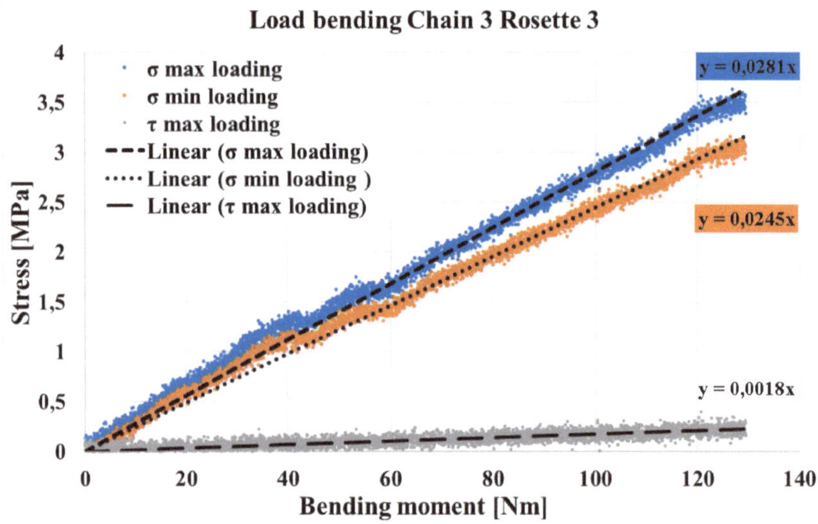

Fig. 3.58 Variation of maximum and minimum normal stresses, and maximum shear stress, for rosette 3 in chain 3 under bending load – loading phase

Figure 3.59, the variation of the three stresses, maximum and minimum principal normal stresses and maximum shear stress, for **bending loading**, at the applied load, corresponding to **rosette 4 in chain 3**, is presented. Rosette 4 is located slightly farther from the blade tip compared to rosette 3, with a distance of 5 mm between the rosettes. The following can be observed:

-The highest stress values occur for the maximum normal stresses, with the largest value being approximately **3.1 MPa** (**5 MPa** was recorded for rosette 1 in chain 3, **3.4 MPa** for rosette 2 in chain 3, and **3.6 MPa** for rosette 3 in chain 3);

-The minimum stress, which is approximately 2.6 MPa, is close to the maximum stress, suggesting an approach to a biaxial state of tensile stresses;

-The shear stress values are close to zero.

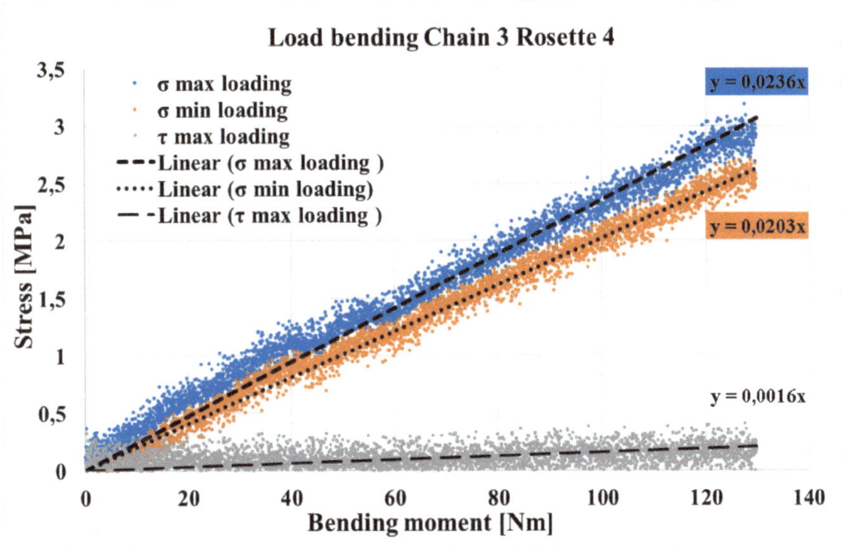

Fig. 3.59 Variation of maximum and minimum normal stresses, and maximum shear stress, for rosette 4 in chain 3 under bending load – loading phase

In Figure 3.60, the variation of the three stresses, maximum and minimum principal normal stresses and maximum shear stress, for **bending loading**, at the applied load, corresponding to **rosette 5 in chain 3**, is presented. Rosette 5 is located slightly farther from the blade tip compared to rosette 4, with a distance of 5 mm between the rosettes. The following can be observed:

- The highest stress values occur for the maximum normal stresses, with the largest value being approximately **3 MPa (5 MPa** was recorded for rosette 1 in chain 3, **3.4 MPa** for rosette 2 in chain 3, **3.6 MPa** for rosette 3 in chain 3, and **3.1 MPa** for rosette 4 in chain 3);
- The minimum stress, which is approximately 2.6 MPa, is close to the maximum stress, suggesting an approach to a biaxial state of tensile stresses;
- The shear stress values are close to zero.

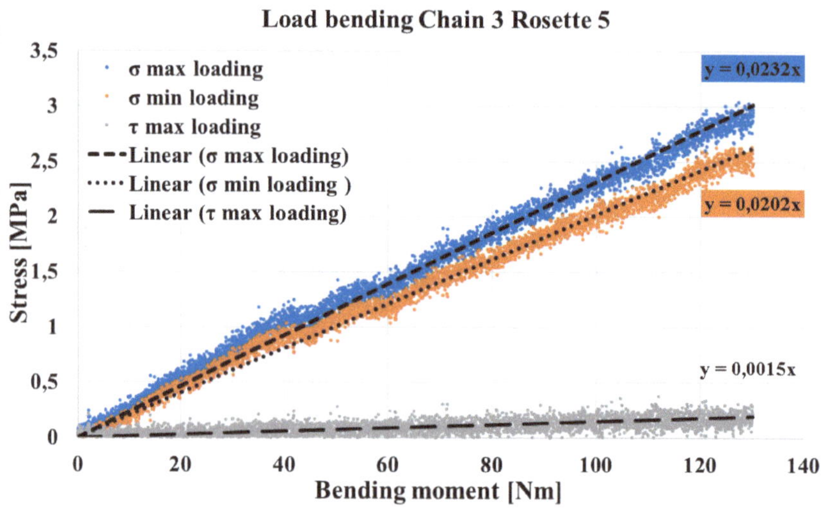

Fig. 3.60 Variation of maximum and minimum normal stresses, and maximum shear stress, for rosette 5 in chain 3 under bending load – loading phase

In Figure 3.61, the variation of the three stresses, maximum and minimum principal normal stresses and maximum shear stress, for **bending loading**, at the applied load, corresponding to **rosette 1 in chain 4**, is presented. As previously mentioned, chain 4 is located 560 mm from the blade tip, where the bending displacement is applied, but on the extrados, where the stresses induced by bending are **compressive**. Chain 4 is in the same cross-section as chain 1, but on the opposite side. The distance between rosettes is 5 mm. The following can be observed:

- The largest stress values occur for the minimum normal stresses, with the largest absolute value being approximately **3.6 MPa** (for rosette 1 in chain 1, we had **3.7 MPa**). The bending moment for chain 4 (considering the slight differences between rosettes), as for chain 1, was approximately 35 Nm;
- The minimum stress, approximately 3.2 MPa, is close to the maximum stress, suggesting an approach to a biaxial state of tensile stresses;
- The shear stress values are close to zero.

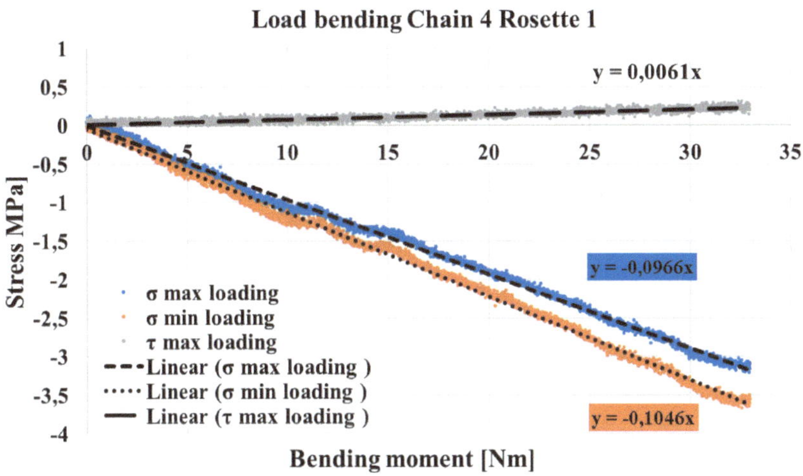

Fig. 3.61 Variation of maximum and minimum normal stresses, and maximum shear stress, for rosette 1 in chain 4 under bending load – loading phase

In Figure 3.62, the variation of the same stresses during **unloading** from bending is presented. Considering that, in all unloading cases, the same trend was observed passing through zero and reaching positive stresses (as long as the stresses during unloading were positive), it can be assumed that, upon reaching zero load, the blade still deforms slightly in the unloading direction, thus resulting in positive stresses. In any case, the stress values during unloading are very close to those during loading.

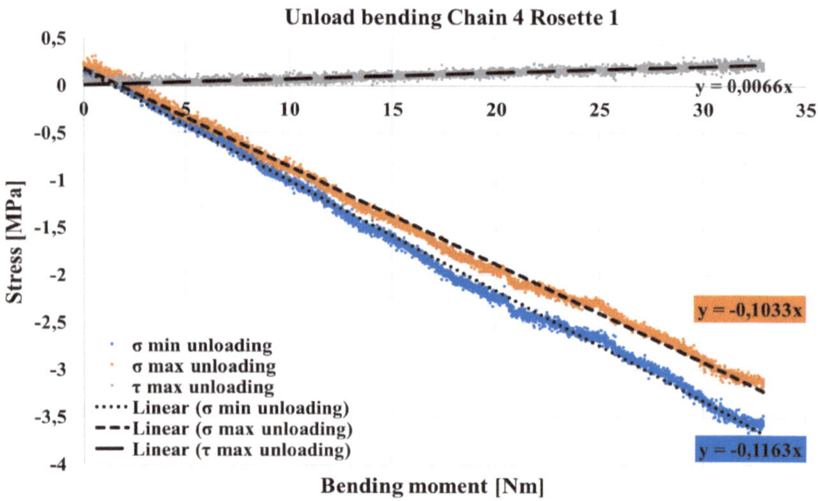

Fig. 3.62 Variation of maximum and minimum normal stresses, and maximum shear stress, for rosette 1 in chain 4 under bending load – loading phase

In Figure 3.63, the variation of the three stresses normal principal, maximum, and minimum, and maximum shear stress for bending under **loading** is presented, corresponding to **rosette 2 in chain 4**. Rosette 2 is located slightly farther from the blade tip compared to rosette 1, with the distance between rosettes being 5 mm. The following observations can be made:

- The highest stress values occur for the maximum normal stresses, with the highest value being approximately **4 MPa** (compared to **4.3 MPa** for rosette 2 in chain 1);
- The minimum stress, approximately 3.8 MPa, is close to the maximum stress, suggesting a nearly biaxial tensile stress state;
- The shear stress values are close to zero.

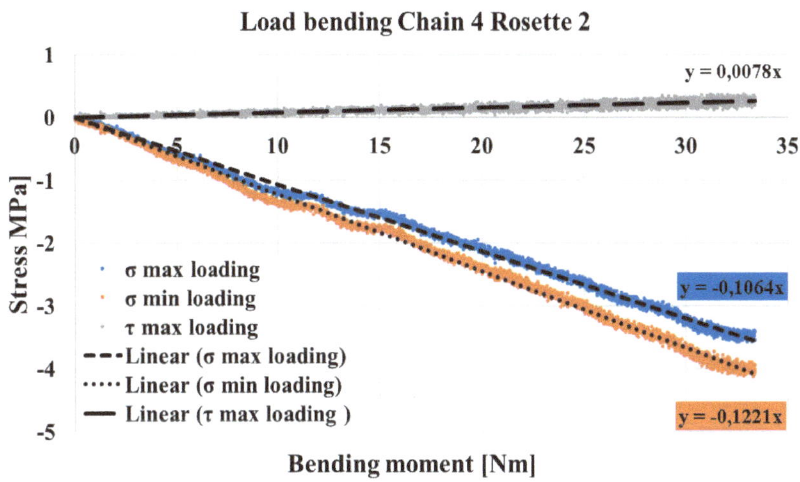

Fig. 3.63 Variation of maximum and minimum normal stresses, and maximum shear stress, for rosette 2 in chain 4 under bending load – loading phase

In Figure 3.64, the variation of the three stresses normal principal, maximum, and minimum, and maximum shear stress for bending under **loading** is presented, corresponding to **rosette 3 in chain 4**. Rosette 3 is located slightly farther from the blade tip compared to rosette 2, with the distance between rosettes being 5 mm. The following observations can be made:

- The highest stress values occur for the maximum normal stresses, with the highest value being approximately **3.9 MPa** (compared to **4.2 MPa** for rosette 3 in chain 1);
- The minimum stress, approximately 3.4 MPa, is close to the maximum stress, suggesting a nearly biaxial tensile stress state;
- The shear stress values are close to zero.

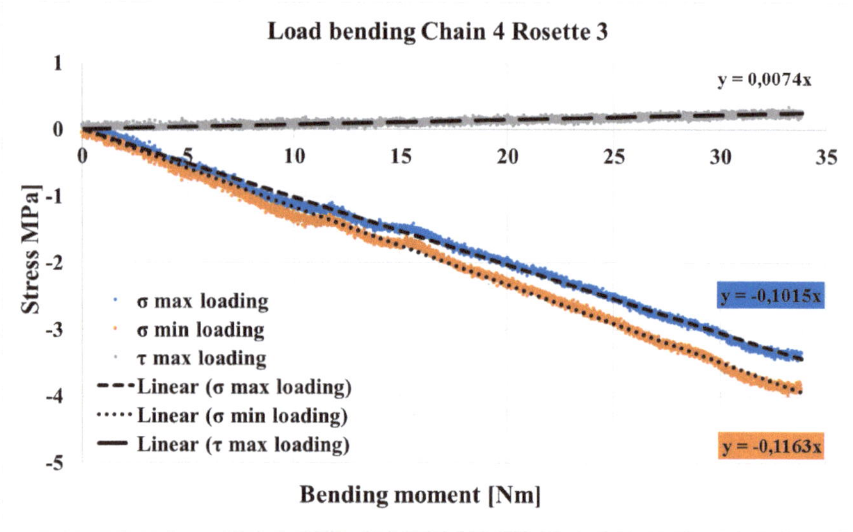

Fig. 3.64 Variation of maximum and minimum normal stresses, and maximum shear stress, for rosette 3 in chain 4 under bending load – loading phase

In Figure 3.65, the variation of the three stresses normal principal, maximum, and minimum, and maximum shear stress for bending under **loading** is presented, corresponding to **rosette 4 in chain 4**. Rosette 4 is located slightly farther from the blade tip compared to rosette 3, with the distance between rosettes being 5 mm. The following observations can be made:

- The highest stress values occur for the maximum normal stresses, with the highest value being approximately **3.5 MPa** (compared to **4.8 MPa** for rosette 4 in chain 1);
- The minimum stress, approximately 3.1 MPa, is close to the maximum stress, suggesting a nearly biaxial tensile stress state;
- The shear stress values are close to zero.

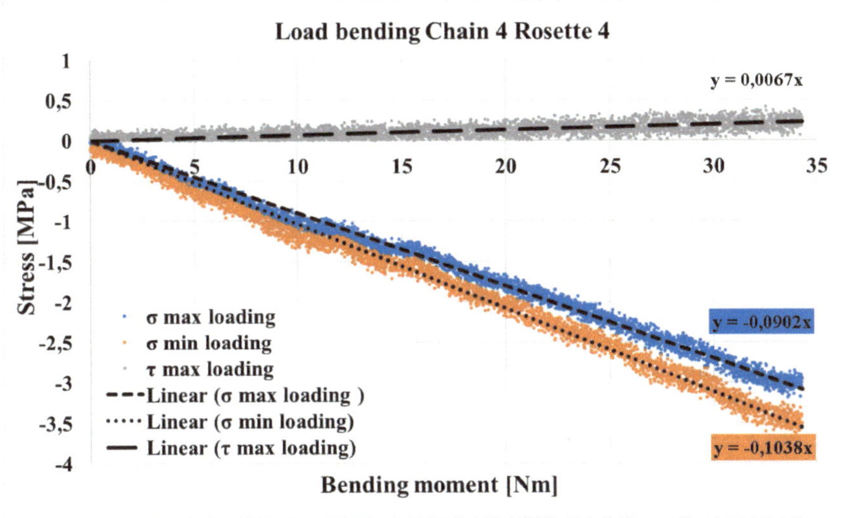

Fig. 3.65 Variation of maximum and minimum normal stresses, and maximum shear stress, for rosette 4 in chain 4 under bending load – loading phase

In Figure 3.66, the variation of the three stresses principal normal, maximum, and minimum, and the maximum shear stress is presented for **bending loading**, corresponding to **rosette 5 in chain 4**.

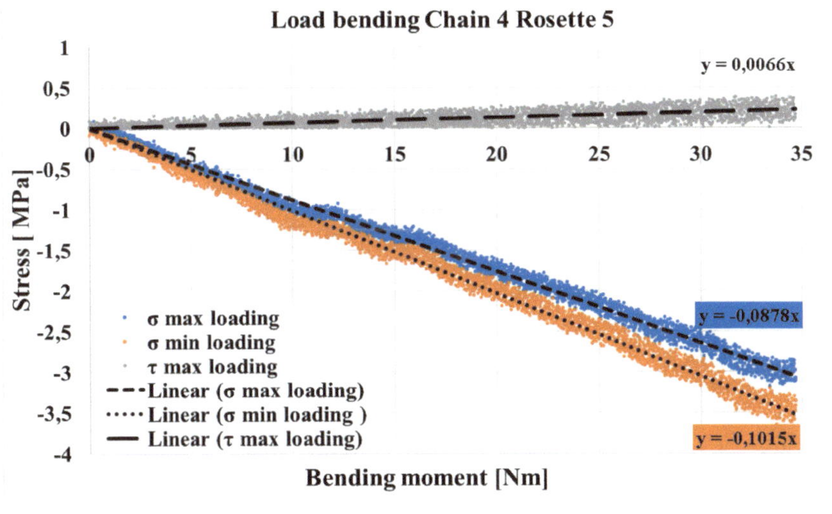

Fig. 3.66 Variation of maximum and minimum normal stresses, and maximum shear stress, for rosette 5 in chain 4 under bending load – loading phase

Rosette 5 is located slightly further from the blade tip compared to rosette 4, with a distance of 5 mm between rosettes. The following can be observed:

- The highest stress values occur for the maximum stresses, with the peak value being approximately **3.5 MPa** (compared to **4.1 MPa** for rosette 5 in chain 1);
- The minimum stress, approximately 3 MPa, is close to the maximum stress value, suggesting an almost biaxial tensile stress state;
- The shear stress values are close to zero.

The following presents Table 3.6, which summarizes the highest values obtained during bending loading for the strain gauges belonging to 4/120 KY31 chains. It can be observed that, for rosettes within the same chain, the maximum values of the same type of stress are close to each other.

For example, for the rosettes in chain 1, the highest values of the maximum principal normal stress are:

Rosette 1	Rosette 2	Rosette 3	Rosette 4	Rosette 5
3.74	4.2	3.92	3.76	4.18

Although rosette 1 is closer to the free end of the blade, where the bending force is applied, a linear increase of σ_{max} toward rosette 5 is not observed (even though the moment arm of the force increases). This occurs because the cross-sectional geometry also changes, with the axial section modulus, W_z, increasing as we move away from the free end of the blade.

Table 3.6 The maximum values of the principal normal stresses (both maximum and minimum) and of the maximum shear stress

Stress [MPa]	**BENDING OF THE ROSETTES IN THE CHAINS**				
	CHAIN 1				
	Rosette 1	**Rosette 2**	**Rosette 3**	**Rosette 4**	**Rosette 5**
σ_{max}	3.74	4.2	3.92	3.76	4.18
σ_{min}	3.3	3.67	3.42	3.3	3.63
τ_{max}	0,41	0.4	0.36	0.42	0.47
	CHAIN 2				
σ_{max}	3.09	2.99	2.88	2.98	3.42
σ_{min}	2.72	2.7	2.53	2.61	2.87
τ_{max}	0.35	0.36	0.33	0.34	0.51
	CHAIN 3				
σ_{max}	5.02	3.42	3.63	3.18	3.04

σ_{min}	4.36	2.99	3.15	2.7	2.63
τ_{max}	0.52	0.41	0.39	0.4	0.37
			CHAIN 4		
σ_{max}	-3.21	-3.57	-3.45	-3.18	-3.17
σ_{min}	-3.67	-4.12	-3.97	-3.62	-3.65
τ_{max}	0.31	0.34	0.34	0.41	0.39

3.7.2 Blade Torsion Loading

3.7.2.1 Clockwise Torsion Loading

The blade at the previously described scale, made of composite material, was subjected to torsion loading in both directions, taking into account that the blade's cross-section varies along its length in both size and shape. The specific strains were recorded from 20 strain gauge transducers, mounted as five on each of the four chains of type 4/120 KY31, manufactured by HBM, shown in Figure 3.5. The strain gauges (strain gauge rosettes) contain three gauges with converging axes, positioned at angles of 0°/120°/240° relative to each other, with a distance of 5 mm between their centers.

The acquisition of specific strains from the 20 transducers, totaling 60 gauges, was performed using four data acquisition systems of type QUANTUM HBM MX1615B. Measurements were conducted dynamically, with a sampling frequency of 20 Hz, and the software interface connection was of the "sg 3 wire 120" type. Data acquisition was carried out simultaneously, with three trials performed under the same conditions. The results presented in this work correspond to a single trial, noting that similar values were obtained across all three trials.

Torsional loading was applied by rotating the free end of the blade using a stepper electric motor, through two gear reducers. Relative to the free end of the blade, where the loading was applied, the strain gauge chains were positioned as follows:

- L1 at a distance of 400 mm on the intrados;
- L2 at a distance of 560 mm on the intrados;
- L3 at a distance of 1500 mm on the intrados;
- L4 at a distance of 400 mm on the extrados.

The transducers L1/L4, L2, and L3 being positioned at different distances, the loading produced the same torsional moment acting on different cross-sections. Moreover, since the distance between the strain gauge rosettes was 5 mm, and from one rosette to another within the same transducer, the shape and dimensions of the cross-section varied, with the cross-sectional area increasing from the free end of the blade (the zone of torsional moment application) toward the clamped end of the blade.

The torsion tests were performed by rotating the blade both clockwise and counterclockwise, with a maximum rotation of 4 degrees corresponding to a torsional moment of 56 Nm, obtained through the previously mentioned calibration. The loading was applied within the elastic range; therefore, the increase of stresses relative to the applied load is expected to be linear. Given the high acquisition frequency (20 Hz) and the transmission of the torsional moment through the gear reducers, in the following graphs, both in the variation of the stresses and of the principal normal stress angle, the following can be observed:

- Some scattering of the stress results around an approximating straight line;
- A few small deviations from the alignment of values on the straight line. It should be noted that these deviations are smaller than in the bending loading case, due to the torsional moment being transmitted via two gear reducers.

In any case, the important aspect is the trend highlighted by the corresponding straight lines, as the loadings were carried out entirely within the elastic range.

Using the calculation formulas presented earlier, the successive values were determined, both for loading and unloading, for:

- Maximum normal stress;
- Minimum normal stress;
- Maximum shear stress;
- Angle of the principal normal stress.

Considering that the obtained values and trends are similar, only a few significant results will be presented below, both for loading and unloading. At the end, the maximum values obtained for each rosette will be presented in a table, for the loading condition.

Figure 3.67 shows the variation of stresses: maximum principal normal stress, minimum principal normal stress, and maximum shear stress, for **rosette 2**

from chain 1, under **clockwise torsion** loading, for the case of loading from zero to approximately 56 Nm. The following can be observed:

- During loading, the stresses increase continuously, suggesting correct data acquisition, except for the scatter mentioned earlier;
- The highest stress values are recorded for the maximum principal normal stresses, with the highest value being approximately 0.75 MPa;
- The minimum stress value is close to the maximum stress value, indicating an approach to a biaxial tensile stress state;
- Because the maximum and minimum normal stresses are close, the shear stress value approaches zero.

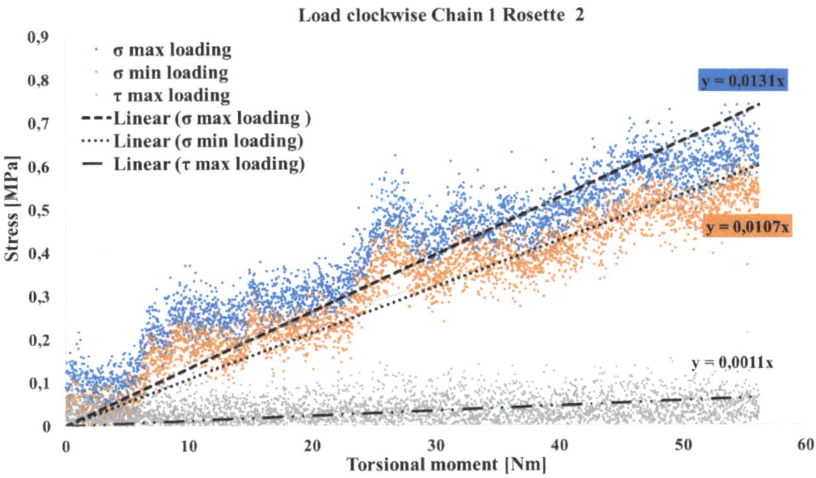

***Fig. 3.67** Variation of maximum and minimum normal stresses, and maximum shear stress, for rosette 2 in chain 1 under clockwise torsion-loading*

In Figure 3.68, the variation of the angle of the maximum principal stress (relative to the direction of gauge A in the rosette) is shown during loading from a torsion moment of zero up to approximately 56 N·m. The values fluctuate around zero degrees, with the oscillations shown in the figure being caused by the way the torque is transmitted by the stepper motor through two gear reducers. The sampling time is relatively short, at 0.05 seconds (20 Hz), resulting in a large number of acquired data points and, consequently, multiple oscillatory values.

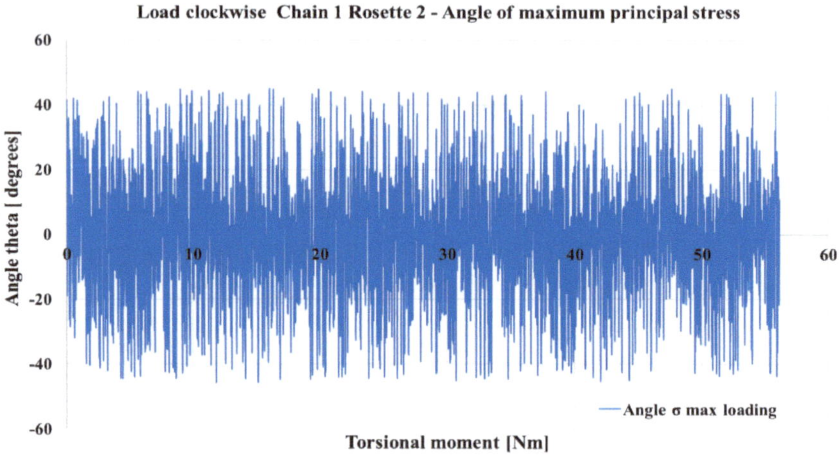

Fig. 3.68 Variation of the maximum principal normal stress angle with the torsion moment, for rosette 2 in chain 1 – clockwise torsion during loading

Figure 3.69 shows the variation of the previously mentioned stresses during **unloading** under torsion. As with bending, it can be observed that the values do not return exactly to zero, and the explanations provided for the bending load are also valid in this case.

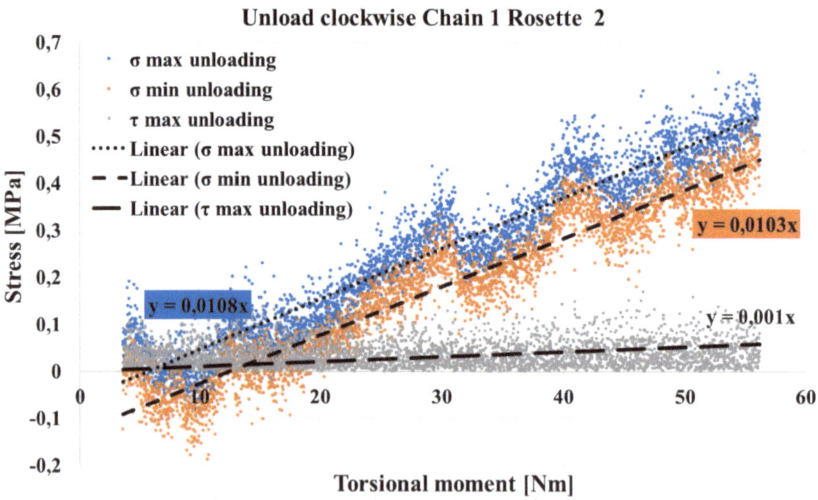

Fig. 3.69 Variation of maximum and minimum normal stresses, and maximum shear stress, for rosette 2 in chain 1 under clockwise torsion-unloading

Figure 3.70 shows the graphs of the variation with torsion moment of the maximum principal normal stress, the minimum principal normal stress, and the maximum shear stress for one of the rosettes belonging to **chain 4**, which is positioned on the extrados (all the other chains are positioned on the intrados).

It can be observed that the minimum and maximum normal stresses are negative, with the largest absolute value of the minimum stress being greater than the largest absolute value of the maximum stress, which explains the positive values of the shear stresses.

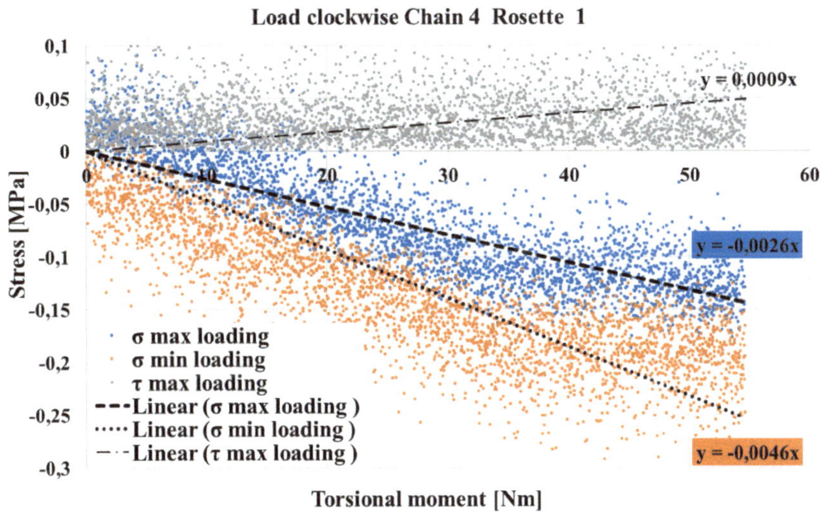

Fig. 3.70 Variation of maximum and minimum normal stresses, and maximum shear stress, for rosette 1 in chain 4 under clockwise torsion- loading

Because the variations of the three stresses are similar, and taking into account the differences in sign between chains L1, L2, L3, and L4, not all the graphs obtained from the tests are presented further. However, in Table 3.7, we present the highest values obtained during clockwise torsion loading for the strain gauge rosettes belonging to the 4/120 KY31 chains. No trend of decreasing stresses is observed as the polar section modulus, W_p, increases and as we move away from the free end of the blade, considering that the torsion moment is constant.

Rosette 1	Rosette 2	Rosette 3	Rosette 4	Rosette 5
0.44	0.79	1.01	0.79	1.23

Table 3.7 Maximum values of the principal normal stresses (both maximum and minimum) and of the maximum shear stress – clockwise torsion loading

STRESS [MPa]	CLOCKWISE TORSION – LOADING OF THE ROSETTES IN THE CHAINS				
	CHAIN 1				
	Rosette 1	**Rosette 2**	**Rosette 3**	**Rosette 4**	**Rosette 5**
σ_{max}	0.44	0.79	1.01	0.79	1.23
σ_{min}	0.31	0.6	0.7	0.67	0.96
τ_{max}	0.21	0.21	0.29	0.16	0.27
	CHAIN 2				
σ_{max}	0.43	0.34	0.29	0.32	0.45
σ_{min}	0.27	0.22	0.16	0.15	0.17
τ_{max}	0.19	0.15	0.17	0.22	0.29
	CHAIN 3				
σ_{max}	0.70	0.48	0.73	0.34	0.83
σ_{min}	0.52	0.33	0.56	0.42	0.68
τ_{max}	0.20	0.16	0.21	0.24	0.24
	CHAIN 4				
σ_{max}	-0.18	-0.15	-0.20	-0.13	0.13
σ_{min}	-0.33	-0.25	-0.32	-0.37	-0.23
τ_{max}	0.14	0.11	0.12	0.25	0.11

3.7.2.2 Anticlockwise torsion loading.

Figure 3.71 shows the variation of the stresses: maximum principal normal, minimum principal normal, and maximum shear stress for **rosette 5 from chain 1**, under **anticlockwise torsion loading**, from zero to approximately 56 Nm. The following observations can be made:

- During the loading process, the stresses vary almost linearly, with deviations whose causes were discussed previously;
- For the normal stresses in chains 1, 2, and 3, the values are negative, with the minimum stress having a greater absolute value than the maximum stress;
- Even under counterclockwise torsion, the stresses are small, the highest absolute value being approximately 0.7 MPa for the minimum principal normal stress;
- The maximum and minimum normal stresses are close in value;

- The highest stress values occur for the maximum principal normal stress, with the greatest value being approximately 0.75 MPa;
- The closeness of the minimum and maximum stresses suggests a biaxial state of compressive stresses;
- Because the maximum and minimum normal stresses are close, the maximum shear stress is nearly zero.

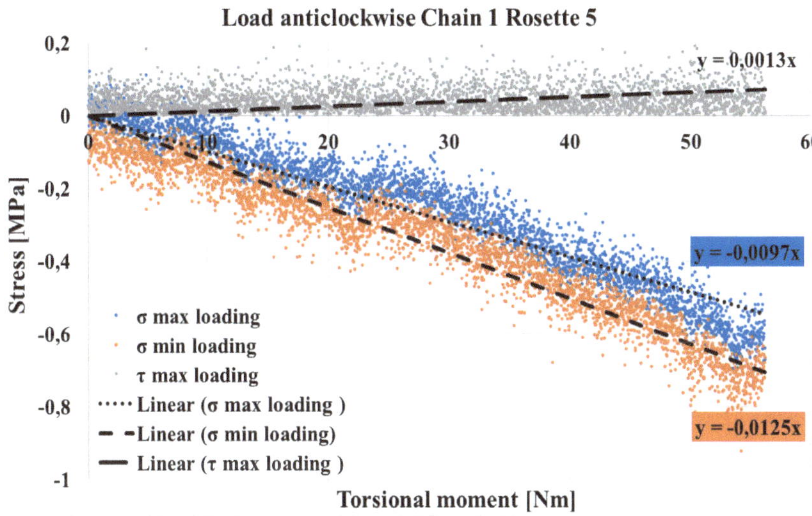

Fig. 3.71 Variation of stresses: maximum normal, minimum normal, and maximum shear, for rosette 5 from chain 1 – anticlockwise torsion – loading

Figure 3.72 shows the variation of the maximum principal normal stress angle (relative to the direction of mark A on the rosette) during loading from zero torsion moment up to approximately 56 N·m. The values fluctuate around zero degrees, with the oscillations shown in the figure resulting from the way the torque is transmitted by the stepper motor through the gear reducers.

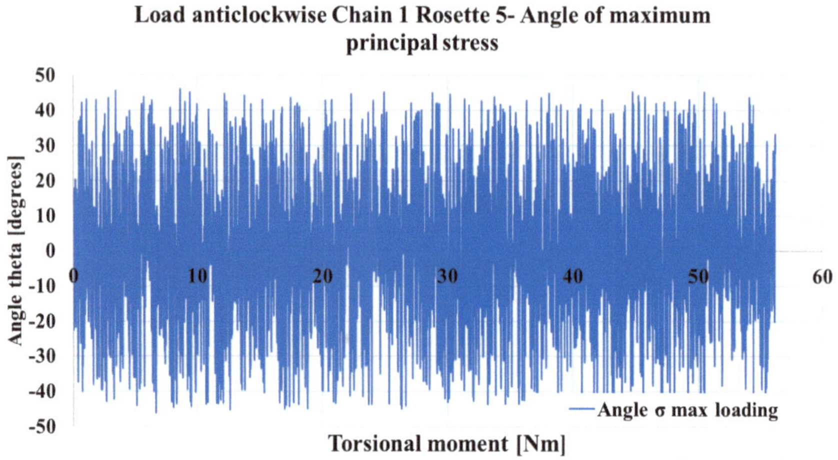

Fig. 3.72 Variation of the maximum principal normal stress angle with the torsion moment for rosette 5 from chain 1 – anticlockwise torsion – loading

Figure 3.73 shows the variation of maximum normal, minimum normal, and maximum shear stresses during **unloading** under torsion for rosette 5 from chain 1, under anticlockwise torsion. It can be observed that the values do not return exactly to zero, with the explanations provided for bending loading being applicable here as well

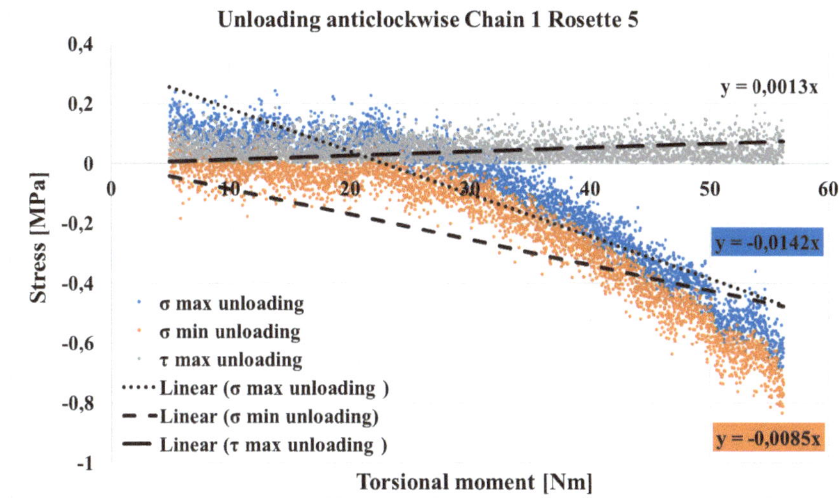

Fig. 3.73 Variation of stresses: maximum normal, minimum normal, and maximum shear, for rosette 5 from chain 1 – anticlockwise torsion – unloading

Figure 3.74 presents the graphs showing the variation with the torsion moment of the maximum principal normal stress, minimum principal normal stress, and maximum shear stress for one of the rosettes belonging to chain 3, which is positioned on the extrados (all other chains are positioned on the intrados). It can be observed that the maximum normal stresses are positive, while the minimum stress values are very close to zero.

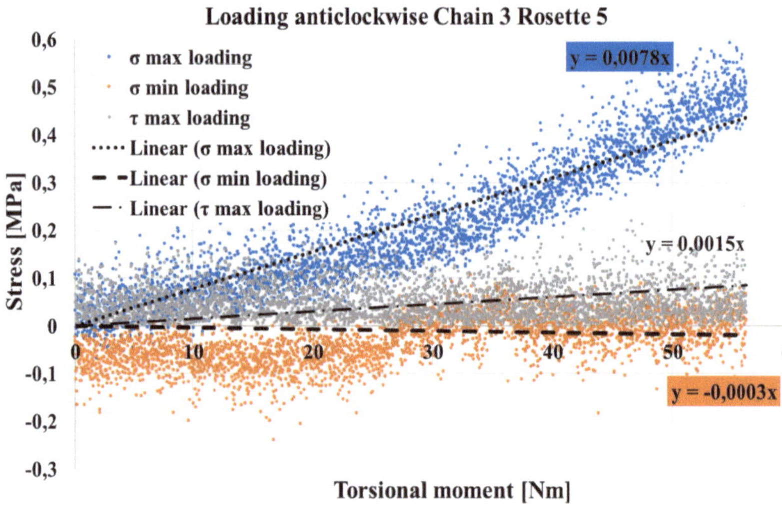

Fig. 3.74 Variation of stresses: maximum normal, minimum normal, and maximum shear, for rosette 5 in chain 3 – anticlockwise torsion – loading

Because the variations of the three stresses are similar, and taking into account the differences in sign between chains L1, L2, L3, and L4, not all graphs obtained from the tests are presented hereafter.

However, Table 3.8 shows the highest values obtained during the torsion loading of the strain gauges belonging to the 4/120 KY31 chains.

No trend of stress reduction is observed as the polar section modulus, W_p, increases that is, as we move away from the free end of the blade given that the applied torsional moment is constant. On the contrary, it is observed that, from rosette 1 to rosette 3, the stress increases (in absolute value), after which, at rosette 4, a decrease in stress is noted.

Rosette 1	Rosette 2	Rosette 3	Rosette 4	Rosette 5
-0.12	-0.22	-0.40	-0.38	-0.69

For the other values, no clear trend in the variation of the principal normal stresses is observed.

Tabel 3.8 Maximum values of the principal normal stresses (both maximum and minimum) and of the maximum shear stress – anticlockwise torsion loading

STRESS [MPa]	ANTICLOCKWISE TORSION – LOADING OF THE ROSETTES IN THE CHAINS				
	CHAIN 1				
	Rosette 1	Rosette 2	Rosette 3	Rosette 4	Rosette 5
σ_{max}	-0.12	-0.22	-0.40	-0.38	-0.69
σ_{min}	-0.26	-0.34	-0.54	-0.53	-0.92
τ_{max}	0.23	0.15	0.16	0.13	0.25
	CHAIN 2				
σ_{max}	3.31	0.18	0.31	0.33	0.25
σ_{min}	-0.21	-0.25	-0.30	-0.18	-0.23
τ_{max}	0.23	0.16	0.24	0.16	0.18
	CHAIN 3				
σ_{max}	-0.75	-0.13	0.15	-0.29	0.60
σ_{min}	-0.88	-0.21	-0.17	-0.53	-0.25
τ_{max}	0.19	0.12	0.16	0.26	0.22
	CHAIN 4				
σ_{max}	-0.09	-0.12	-0.13	-0.09	-0.05
σ_{min}	-0.22	-0.19	-0.22	-0.32	-0.18
τ_{max}	0.11	0.09	0.09	0.24	0.15

3.7.3 Compound load of bending and torsion of the wind turbine blade

Using the experimental setup, it is possible to apply both bending and torsion loads to the scaled blade made of GFRP material. The two types of loading are applied as follows:

- Bending load is applied by translating the free end of the blade in the vertical direction (intrados–extrados) using a stepper motor and an additional associated mechanism;
- Torsional load is applied in both clockwise and anticlockwise directions by rotating the free end of the blade with the help of a stepper motor and two gear reducers. To prevent rotation of the entire actuation assembly, a guiding system is provided, which simultaneously allows both vertical translation (producing bending) and rotation (in either direction).

The four chains of transducers, each with 5 rosettes having three grids positioned at 0°/120°/240°, are the same as those previously described. Data from QUANTUM HBM MX1615B strain gauge bridges were acquired at intervals of

1/20 seconds, with three tests conducted under the same conditions. Only one of the three tests, which is more illustrative, is presented.

3.7.3.1 Bending with clockwise torsion loading

This type of loading was applied by simultaneously translating and rotating the free end of the blade in the clockwise direction. The placement of the strain gauge chains is as follows: L1, L2, and L3 are positioned on the intrados, where bending induces tensile stresses, while L4 is positioned on the extrados, where bending induces compressive stresses. Relative to the free end of the blade, where the loads are applied, the chains are positioned as follows: L1 (intrados) and L4 (extrados) are located 400 mm from the free end, L2 is at 560 mm, and L3 is at 1500 mm. The distance between rosettes is 5 mm, meaning that each rosette experiences a different lever arm and, consequently, a different bending moment. The torsion moment is the same for all rosettes: a 4° rotation of the blade tip produces a torsion moment of approximately 56 N·m. At the maximum tip displacement of 15 mm, a force of approximately 86 N is applied, generating bending in the intrados–extrados direction. These force and torsion values were determined previously through calibration with weights, as described in an earlier section. Moreover, it should be noted that both the axial section modulus W_z and the polar section modulus W_p vary from one rosette to another, due to the blade geometry, which increases in size toward the clamped root.

Considering the short data acquisition interval (0.05 s) as well as the oscillations caused by the discontinuities in the transmission through the gear reducers, the following graphs will show, for both the stresses and the angle of the maximum principal stress:

- A certain scatter of the results at stress;
- Oscillation around a trend at angle.

Using the calculation formulas presented earlier, the successive values were determined, both for loading and unloading, for:

- Maximum normal stress;
- Minimum normal stress;
- Maximum shear stress;
- Angle of the principal normal stress.

The results obtained are represented through graphs showing a linear trend of variation, taking into account that the loading remained within the elastic domain. Since the shapes of the graphs are similar and the obtained values are close, only a ***few of the resulting graphs are presented here***. In any case, at the end of the paragraph, the maximum stress values obtained for each individual test are presented in a table.

On the abscissa axis is given the compound moment, M_c, which resulted as follows:

$$M_c = \sqrt{M_i^2 + M_t^2}$$

where:

- M_i is the bending moment calculated as the force (obtained by calibration for any moment of displacement) multiplied by the arm up to the respective rosette;
- M_t is the torsional moment, also calculated by calibration for any moment of rotation.

Figure 3.75 shows the variation of stresses maximum normal, minimum normal, and maximum shear for rosette 1 in chain 1, introduced in the blade under **compound bending and clockwise torsion** during the loading phase.

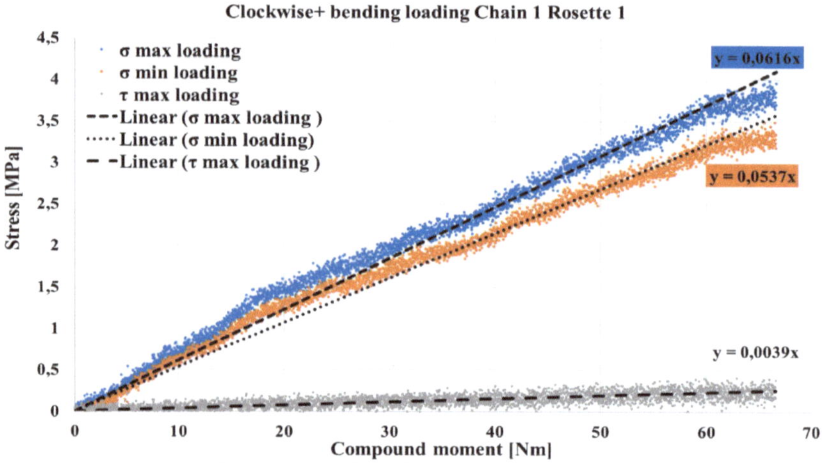

Fig. 3.75 Variation of stresses: maximum normal, minimum normal, and maximum shear—for rosette 1 in chain 1 under compound bending and clockwise torsion loading

From this figure, the following observations can be made:

- Considering the normal stress values obtained at the end of the test, both for bending and torsion separately, it is observed that the stresses induced by bending are predominant;
- The highest stress values are recorded for the maximum principal normal stresses, with the largest value measured at the end of loading being approximately 4 MPa;
- *Compared to simple bending load, the maximum values of the maximum principal normal stress are higher by approximately 0.5 MPa;*
- The minimum stress values are close to the maximum stress values, suggesting a state approaching biaxial tensile stress, given that both minimum and maximum stresses are positive;
- Because the maximum and minimum normal stress values are close, the shear stress approaches zero.

Figure 3.76 shows the variation of the angle of the maximum principal stress (relative to the direction of mark A on the rosette—see Figure 3.5) during loading from zero torsion moment up to approximately 56 N·m, simultaneously with bending loading from zero up to a bending moment of approximately 32 N·m. The values obtained from the calculations oscillate around zero degrees, with the oscillations shown in the figure resulting from the transmission mechanism, using the two motors through the two gear reducers. As previously mentioned, the data acquisition sampling time was small, namely 0.05 seconds. Larger variations are observed at lower loads, indicating that the load transmission chain through the gear reducers exhibits discontinuities.

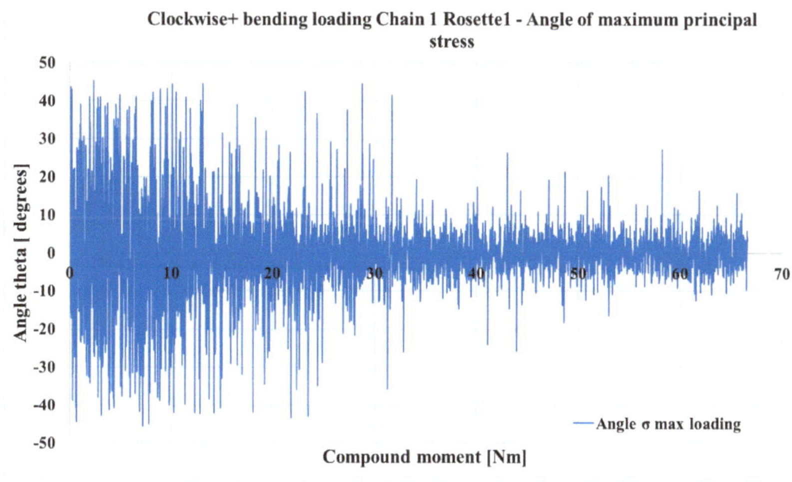

a) Chain 1, rosette 1 (Mc≈65 Nm)

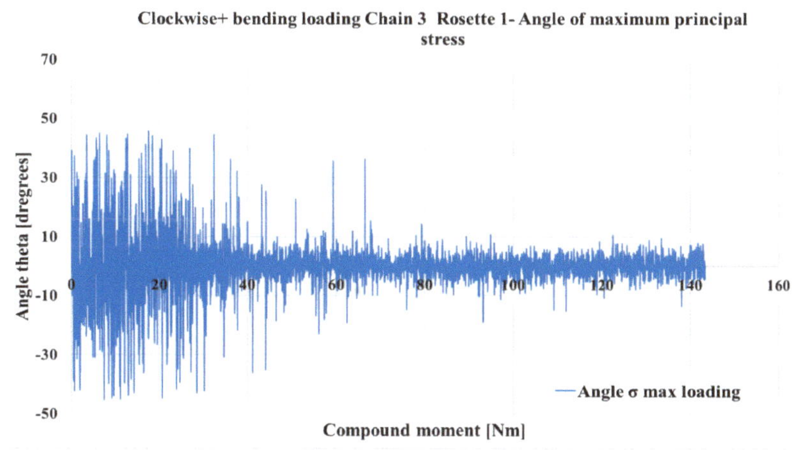

b) Chain 3, rosette 1 (Mc≈145 Nm)

Fig 3.76 Variation of the maximum principal stress angle with the compound moment, for the compound bending and clockwise torsion loading – loading phase

In Figure 3.77, the variation of maximum normal stress, minimum normal stress, and maximum shear stress is presented for rosette 1 in chain 4, installed on the blade under compound bending and clockwise torsion during the loading phase. It should be noted that the rosettes in chain 4 are located in the compressive zone of the blade with respect to the bending load. The following observations can be made from the figure:

- Compared to simple bending loading, the maximum absolute values of minimum normal stress are higher under combined loading, by approximately 0.5 MPa;
- Considering the normal stress values at the end of the test, for both bending and torsion loads taken separately, stresses due to bending are predominant;
- The highest stress values occur for maximum principal stresses, with the peak value at the end of loading being approximately 3.8 MPa;
- Comparing rosettes 1, one on chain 1 (tension side) and the other on chain 4 (compression side), higher normal stress values are obtained on the tension side rosette (4 MPa vs. 3.9 MPa);
- Minimum stress values are close to maximum stress values, suggesting an almost biaxial tensile stress state, since both minimum and maximum stresses are positive;
- Because maximum and minimum normal stresses are close, the shear stress value approaches zero.

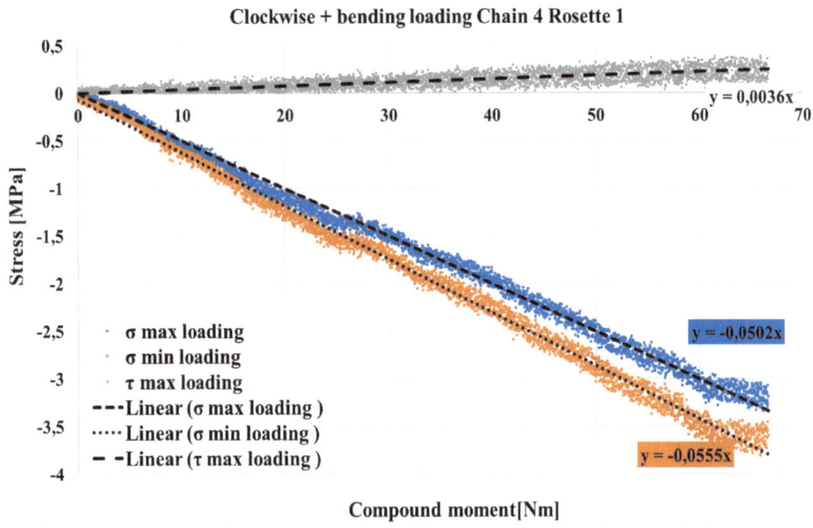

Fig. 3.77 Variation of stresses: maximum normal, minimum normal, and maximum shear, for rosette 1 in chain 4 – compound bending and clockwise torsion – loading phase

In Figure 3.78, the variation of the previously mentioned stresses during **unloading** under the combined bending and torsion load is presented.

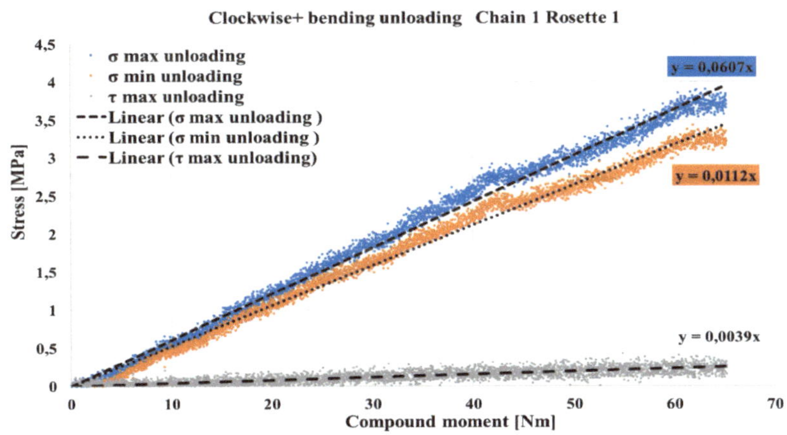

a) Chain 1, rosette 1

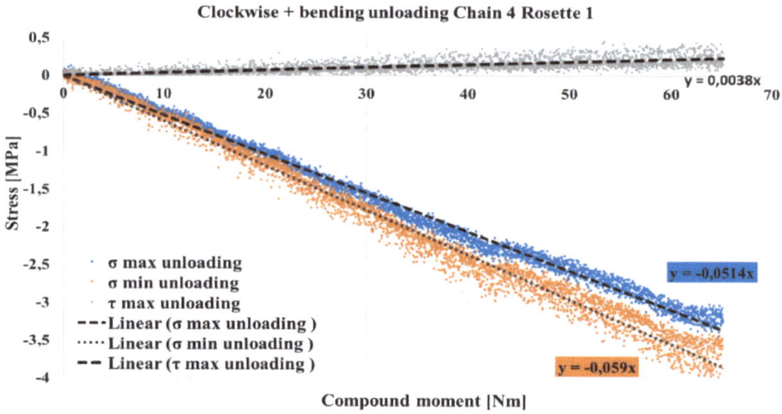

b) Chain 4 rosette 1

Fig. 3.78 Variation of stresses: maximum normal, minimum normal, and maximum shear, for rosette 1 from chains 1 and 4 – compound bending and torsion load in the clockwise direction – unloading

The values are shown for rosettes 1, located on chains 1 and 4. Unlike the previous cases with simple bending or torsion (in either direction), here it is observed that the stress values return to zero as the bending and torsion loads are removed. Although graphs for the other rosettes are not shown (here only two rosettes are presented, one on the extrados and one on the intrados), in all cases it is observed that after unloading from the combined bending and torsion, the stress values return to zero.

Table 3.9 provides the stress values recorded at the end of loading, i.e., under maximum loading, where the compound moment is:

- For chain 1 (average value – varies from rosette to rosette): 68 N·m;
- For chain 2: 76 N·m;
- For chain 3: 141 N·m;
- For chain 4: 68 N·m.

For comparison, the values obtained for the rosettes within chain 1 under **bending loading** are presented below. It can be observed that the maximum principal normal stresses recorded for all rosettes in chain 1, under combined bending and torsion loading, are higher than those obtained under bending alone. This clearly indicates that the tensile stresses have accumulated, even though the contribution of the torsion loading is very small, as was also observed in the simple torsion tests (in both directions).

Bending – loading

Chain 1	Rosette 1	Rosette 2	Rosette 3	Rosette 4	Rosette 5
σ_{max}	3.74	4.2	3.92	3.76	4.18

Table 3.9 Maximum stress values (both principal normal and shear) for the compound bending and torsion loading (clockwise) – loading

STRESS [MPa]	CLOCKWISE TORSION + BENDING – LOADING, FOR THE ROSETTES IN THE CHAINS				
	CHAIN 1				
	Rosette 1	**Rosette 2**	**Rosette 3**	**Rosette 4**	**Rosette 5**
σ_{max}	3.98	4.77	4.75	4.34	5.06
σ_{min}	3.49	4.17	4.14	3.80	4.46
τ_{max}	0.41	0.45	0.49	0.37	0.52
	CHAIN 2				
σ_{max}	3.35	3.14	2.99	3.04	3.49
σ_{min}	2.97	2.76	2.56	2.65	3.05
τ_{max}	0.41	0.29	0.31	0.28	0.39
	CHAIN 3				
σ_{max}	5.29	3.47	4.04	3.19	3.64
σ_{min}	4.60	3.03	3.56	2.17	3.18
τ_{max}	0.52	0.34	0.39	0.56	0.43
	CHAIN 4				
σ_{max}	-3.34	-3.67	-3.54	-3.13	-3.05
σ_{min}	-3.81	-4.18	-4.06	-3.63	-3.52
τ_{max}	0.38	0.41	0.35	0.45	0.31

Figure 3.79 shows the variation of the maximum principal normal stresses for the five rosettes in chain 1, under **bending** alone and under the **compound bending with clockwise torsion**, during the loading phase.

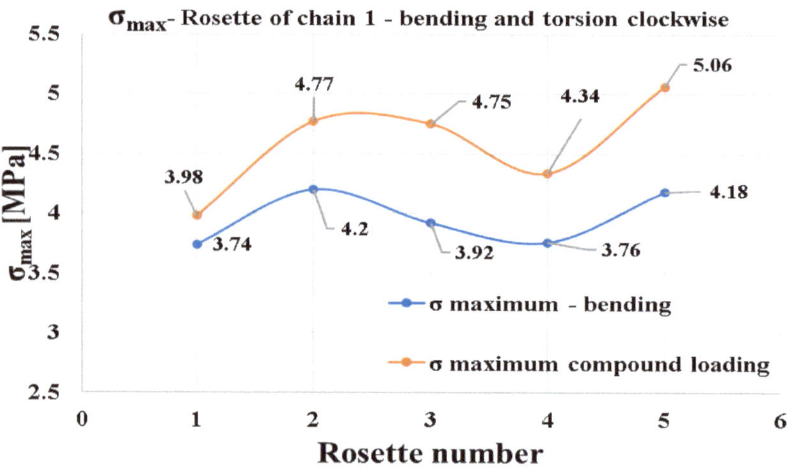

Fig. 3.79 Variation of the maximum principal normal stresses for the rosettes in chain 1, under bending and under the compound bending with clockwise torsion – loading phase

It can be observed that, although the loads occurred at different points in time, and considering that the contribution of stresses from the torsion load is small, the maximum principal normal stresses exhibit the same type of variation in both loading cases.

3.7.3.2 Bending with anticlockwise torsion loading

The anticlockwise torsion load was applied by rotating the free end of the blade in the opposite direction to the previous case, using an electric motor and two gear reducers. The bending load was applied by vertically translating the free end of the blade in the intrados–extrados direction. This displacement was achieved using an electric motor and a vertical guide that allows only the blade to rotate, not the entire actuation assembly. Data acquisition specifically of the strain values was carried out at intervals of 0.05 seconds from the 60 strain gauges, arranged in groups of three per strain gauge rosette, oriented at 0°/120°/240°. The rosettes were factory-mounted in groups of five on a support we refer to as a "rosette chain." The positioning of the chains was presented earlier: L1, L2, and L3 are located on the intrados, where bending produces tensile stresses, while L4 is placed on the extrados, where bending produces compressive stresses. With respect to the free end of the blade—where the loads are applied—the chains are positioned as follows: L1 (intrados) and L4 (extrados) are located at a distance of 400 mm, L2 is positioned at 560 mm, and L3 is positioned at 1500 mm.

The compound bending–torsion loading was applied simultaneously, with the control software for the two stepper motors ensuring both motors started and stopped at the same time. Thus, within the same time interval, a vertical displacement of 15 mm and a rotation (in this case counterclockwise) of 4 degrees were applied. For each moment in time, using the calibration constants obtained and presented earlier for each type of test (bending and torsion), the resulting combined moment was calculated using the relation:

On the abscissa axis, it is given the compound moment, M_c, which resulted as it follows:

$$M_c = \sqrt{M_i^2 + M_t^2}$$

where:

- M_i is the bending moment calculated as the force (obtained by calibration for any moment of displacement) multiplied by the arm up to the respective rosette;
- M_t is the torsional moment, also calculated by calibration for any moment of rotation.

Thus, on the abscissa, in order to maintain the relationship between the applied stresses and the loading, the values of the compound moment were considered representative of the external load. Clearly, from the standpoint of analytical calculations, there is a certain discrepancy between the stress results obtained and those that would be obtained if the combined moment calculated here was used. However, since the stresses were obtained directly from the measured strains, and given that the bending and torsional moments were known for each acquisition point, it is relevant to establish the connection between the stresses and the combined external load, particularly since the stresses are calculated at the surface, where the shear stress produced by the bending force is zero.

While the torsional moment is constant along the entire length of the blade, being applied at the free end, the bending moment varies due to the change in the force arm for each strain gauge. A small error is introduced here, as we considered the lever arm to be the distance from the free end of the blade to the center of the rosette. In reality, as it can be seen in Figure 3.5, each gauge within the rosette is positioned approximately 2–3 mm to either side of the rosette's center. It should also be noted that both the axial section modulus, W_z and the polar section modulus, W_p, vary from one gauge to another due to changes in both the shape and the dimensions of the blade toward its clamped end.

Using the calculation formulas presented earlier, the successive values were determined, both for loading and unloading, for:

- Maximum normal stress;
- Minimum normal stress;
- Maximum shear stress;
- Angle of the principal normal stress.

The results obtained were represented using graphs that exhibited a linear variation trend, given that the loading remained within the elastic range. Since the shapes of the graphs are similar and the values obtained are close to each other, ***only a selection of the graphs will be further presented***. In any case, at the end of the paragraph, the maximum stress values obtained for each individual test are provided in tabular form.

In Figure 3.80, the variation of the maximum normal stress, minimum normal stress, and maximum shear stress is shown for gauge **rosette 1 from chain 1**, embedded in the blade, under compound loading of bending with anticlockwise torsion during the loading phase. From the figure, the following observations can be made:

- Throughout the loading process, the stresses increase continuously, following a linear variation trend, given that the blade was loaded within its elastic range;
- Considering the normal stress values obtained at the end of the test, both for pure bending and for pure counterclockwise torsion, it can be noted that the stresses induced by bending predominate;
- The highest stress values are recorded for the maximum principal normal stresses, the largest of which reaches *approximately 4 MPa* at the end of loading;
- Compared to pure bending simple, the maximum values of the principal normal stresses are about 0.5 MPa higher;

- The minimum stress values are close to the maximum stress values, which suggests a biaxial tensile stress state, given that both minimum and maximum stresses are positive;
- Because the maximum and minimum normal stress values are close, the shear stress value tends toward zero.

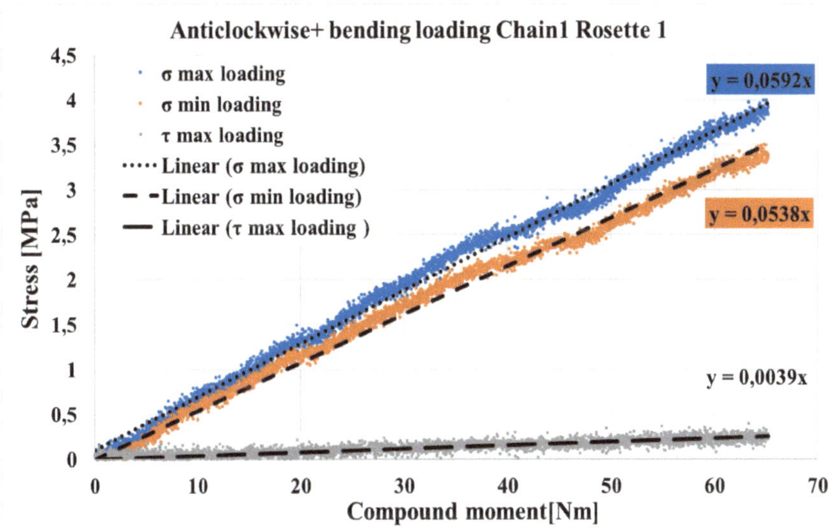

Fig. 3.80 Variation of stresses: maximum normal, minimum normal, and maximum shear, for strain gauge rosette 1 in chain 1 – compound bending and anticlockwise torsion load – loading phase

In Figure 3.81, the variation of the angle of the maximum principal normal stress (relative to the direction of gauge A in the rosette – see Figure 3.5) is shown during loading, from a torsional moment of zero up to the maximum torsional moment, simultaneously with bending from zero up to the maximum bending moment.

Figure 3.81 presents two examples from different chains and different rosettes: **rosette 1 from chain 1 and rosette 4 from chain 2**.

The calculated values oscillate around zero degrees, with the oscillations shown in the figure being caused by the transmission system with the two motors through the two gear reducers. As mentioned earlier, the sampling time for the acquired data was short, namely 0.05 seconds. Larger variations occur at lower loads, confirming that the load transmission system through the gear reducers exhibits discontinuities when the loads are small.

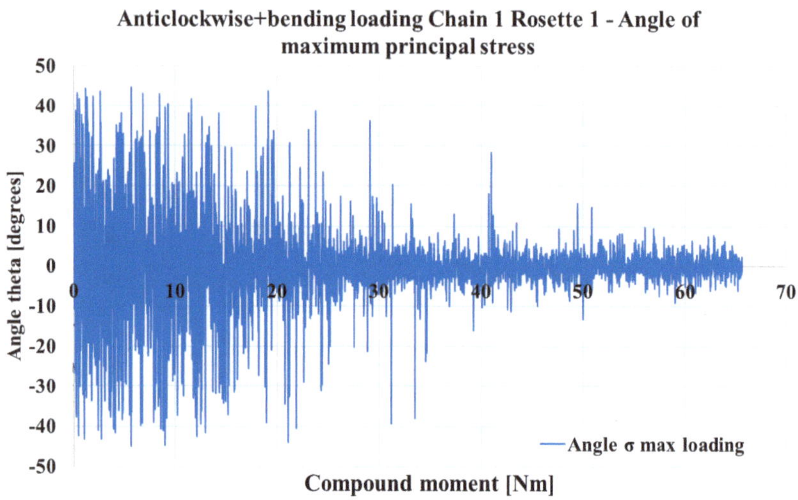

a) Chain 1, rosette 1 (Mc≈66Nm)

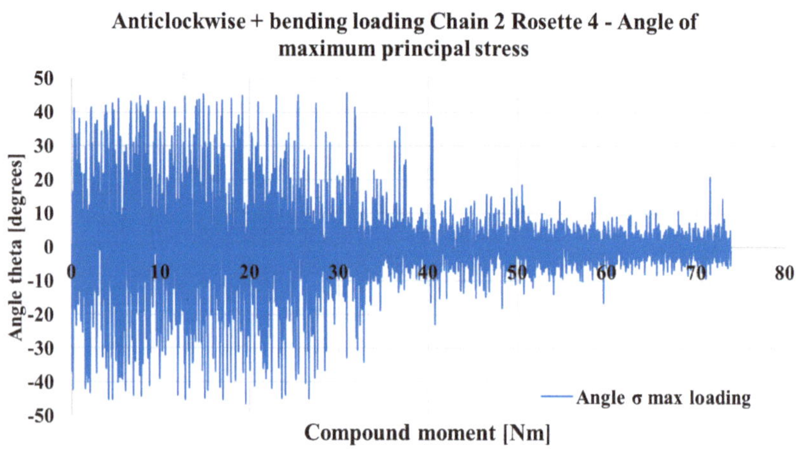

b) Chain 2, rosette 4 (Mc≈74 Nm)

Fig. 3.81 Variation of the maximum principal stress angle with the compound moment, for the compound bending and anticlockwise torsion loading – loading phase

In Figure 3.82, the variation of the maximum normal stress, minimum normal stress, and maximum shear stress is shown for **rosette 1 in chain 4**, introduced into the blade through compound bending with anticlockwise torsion during the loading phase. It is noted that the rosettes in chain 4 are located in the compression zone of the blade with respect to the bending load. From the mentioned figure, the following observations can be made:

- During loading, the stresses increase continuously, which suggests correct data acquisition, except for minor deviations from the linear trend, for reasons mentioned earlier;
- Compared to simple bending, the absolute values of the maximum minimum normal stress are higher for the combined loading, by approximately 0.5 MPa;
- Considering the normal stress values obtained at the end of the test, both for bending and for torsion (separately), it is clear that the stresses developed as a result of bending are predominant;
- The highest stress values are recorded for the maximum principal normal stresses, with the largest value at the end of loading being approximately 3.6 MPa;
- Comparing rosette 1 from chain 1 (upper side – tension) with rosette 1 from chain 4 (lower side – compression), higher normal stress values are observed for rosette 1 on the tension side (4 MPa compared to 3.6 MPa);
- Here as well, the minimum stress values are close to the maximum stress values, suggesting an approach toward a biaxial tensile stress state, considering that both minimum and maximum stresses are positive;

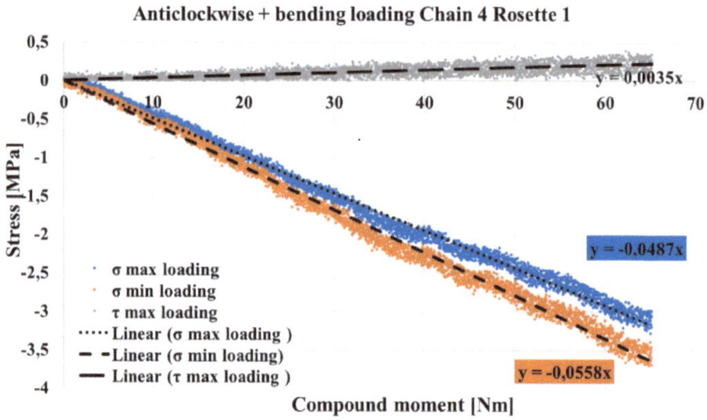

Fig. 3.82 Variation of stresses: maximum normal, minimum normal, and maximum shear, for strain gauge rosette 1 in chain 4 – compound bending and anticlockwise torsion load – loading phase

In Figure 3.83, the variation of the previously mentioned stresses is shown during **unloading** under compound bending and torsion. The values are given for rosette 1, located on chains 1 and 4, similar to the results from the previous figures but for the loading phase. Unlike the earlier cases of simple bending and torsion loading (in either direction), it is observed here that the stress values return to zero during the unloading of the bending and torsional loads. Although graphs for the other rosettes are not presented (here, only two rosettes are shown—one from the extrados side and the other from the intrados side), it is noted that in all cases, after unloading from the combined bending and torsion loading, the stress values return to zero.

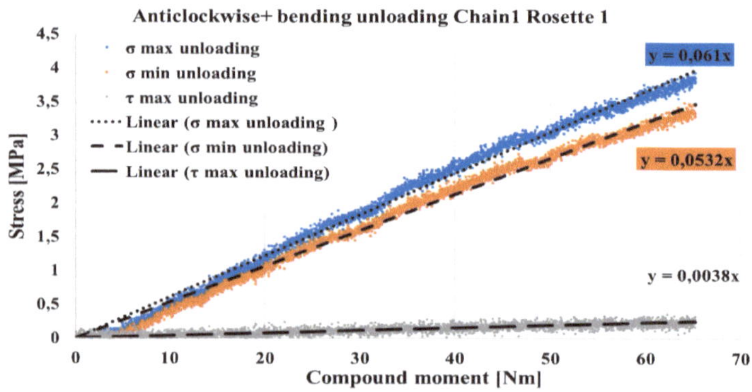

a) Chain 1, rosette 1 – unloading

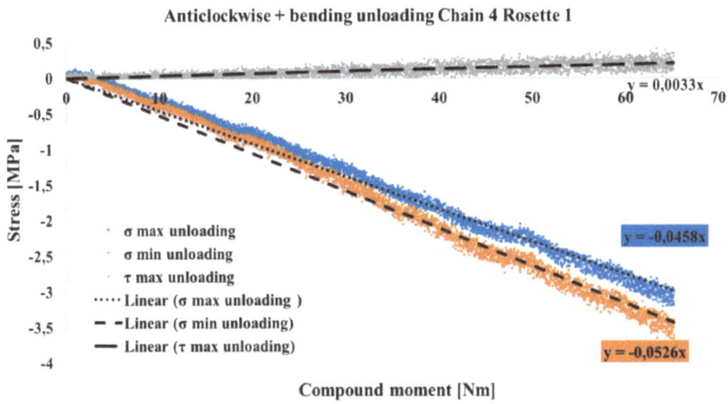

b) Chain 4, rosette 1 – unloading

Fig. 3.83 Variation of maximum normal stress, minimum normal stress, and maximum shear stress for rosette 1 from chains 1 and 4 – compound bending with anticlockwise torsion loading – unloading

Table 3.10 provides the stress values obtained at the end of loading, i.e., under maximum applied loading, where the combined moment is:

- For chain 1 (an average value – varies from rosette to rosette): 68 N·m;
- For chain 2: 76 N·m;
- For chain 3: 141 N·m;
- For chain 4: 68 N·m.

For comparison, the values obtained for the rosettes within chain 1 under **bending loading** are presented below. Unlike the compound loading with clockwise torsion, the trend where all values obtained under combined loading are higher than those for bending alone is no longer maintained. The orientation of the cross-sectional shape relative to the direction of the bending moment, in relation to the shape and dimensions of the neighboring cross-sections, leads to the variations shown in Figure 3.84.

Bending-loading

Chain 1	**Rosette 1**	**Rosette 2**	**Rosette 3**	**Rosette 4**	**Rosette 5**
σ_{max}	3.74	4.2	3.92	3.76	4.18

Tabel 3.10 Maximum stress values (principal normal and shear) for the compound bending and torsion (anticlockwise) loading – loading phase

STRESS [MPa]	**ANTICLOCKWISE TORSION + BENDING – LOADING, FOR THE ROSETTES IN THE CHAINS**				
	CHAIN 1				
	Rosette 1	**Rosette 2**	**Rosette 3**	**Rosette 4**	**Rosette 5**
σ_{max}	3.99	4.28	3.79	3.69	3.45
σ_{min}	3.49	3.73	3.26	3.18	3.04
τ_{max}	0.39	0.46	0.40	0.48	0.42
	CHAIN 2				
σ_{max}	3.19	3.25	3.02	3.33	3.57
σ_{min}	2.84	2.82	2.67	2.91	3.03
τ_{max}	0.36	0.38	0.30	0.44	0.48
	CHAIN 3				
σ_{max}	4.90	3.27	3.13	3.32	2.48
σ_{min}	4.24	2.79	2.73	2.89	2.17
τ_{max}	0.52	0.30	0.33	0.42	0.28
	CHAIN 4				
σ_{max}	-3.26	-3.57	-14.16	-3.16	-3.08
σ_{min}	-3.69	-4.08	-14.62	-3.66	-3.54
τ_{max}	0.39	0.39	0.40	0.47	0.32

Figure 3.84 shows the variation of the maximum principal normal stresses for the five rosettes in chain 1 under **bending** and **compound bending with anticlockwise torsion** during the loading phase. It can be observed that the first two values are higher for the maximum principal normal stresses obtained under the combined loading, while the next three values are higher in the case of bending loading.

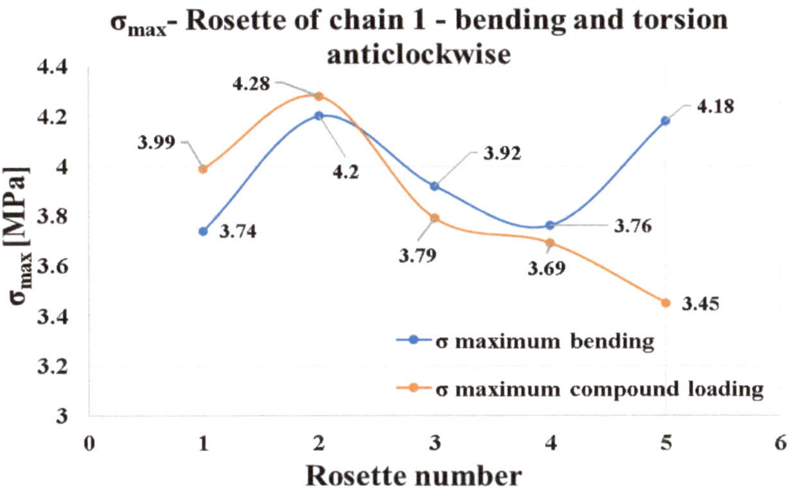

Fig. 3.84 Variation of the maximum principal normal stresses for the rosettes in chain 1, under bending and under the compound bending with anticlockwise torsion – loading phase

3.8 Validation of experimental results with FEM

The simulations presented in Chapter 2 were necessary to identify the regions with maximum stresses, where the rectangular strain gauges and the four delta-type strain gauge chains were mounted. Following the calibrations, a maximum displacement of 15 mm was obtained under bending load, which corresponded to a maximum force of approximately 86 N. The maximum rotation reached 4 degrees, resulting in a maximum torsional moment of approximately 56 N·m. These maximum force and torsional moment values were used as reference for the FEM analyses.

For the five loading cases, the equivalent von Mises stresses and the maximum principal stress are presented below.

The relative errors between the maximum von Mises stresses determined experimentally and those obtained via FEM were calculated for the entire blade. Although the von Mises criterion is primarily used for metallic materials, its use was justified based on the following considerations:

- The composite material was considered homogeneous and orthotropic in the FEM analysis;
- The Tsai-Hill criterion, used for composite materials, degenerates into the von Mises criterion when the allowable tensile and compressive stresses are equal (Morăraş et al., 2023);
- The ANSYS program provides von Mises stresses;
- The blade was tested within the elastic range, with equivalent stresses far from the failure limit;
- For the tested composite material, the tensile and compressive failure stresses are relatively close.

- Bending load

Figure 3.85 presents the von Mises stress distribution, $\sigma_{ech\ max}$=7.09 MPa

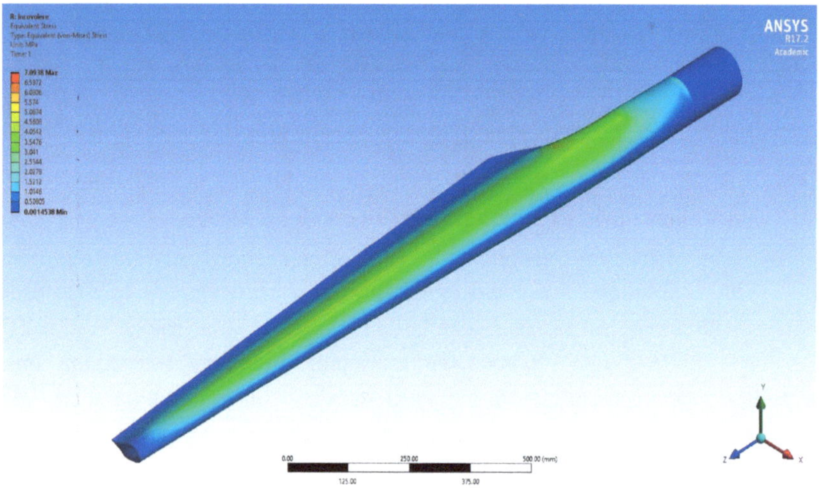

Fig 3.85 Distribution of equivalent von Mises stresses – isometric view, intrados–extrados

Figure 3.86 presents the maximum principal stress distribution, $\sigma_{1\ max}$ =4.92 MPa.

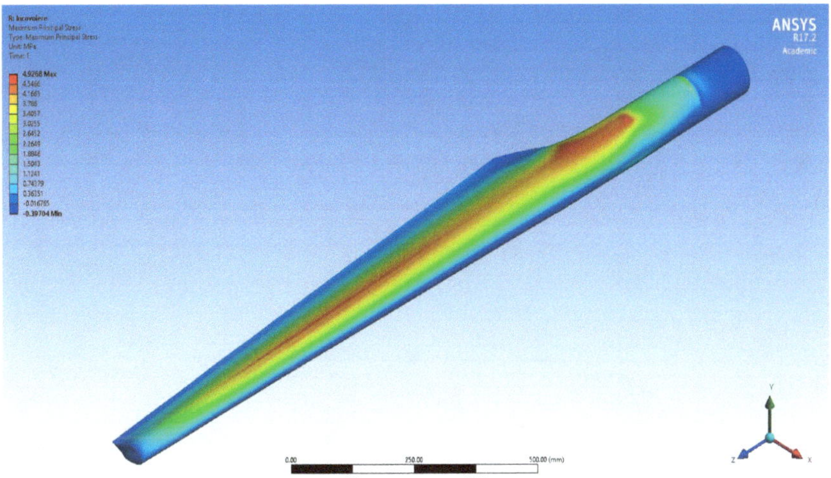

Fig 3.86 Maximum principal stress, isometric view along the intrados–extrados direction

- Torsion loading – clockwise direction

Figure 3.87 presents the von Mises stress distribution, $\sigma_{ech\ max}$=33.88 MPa.

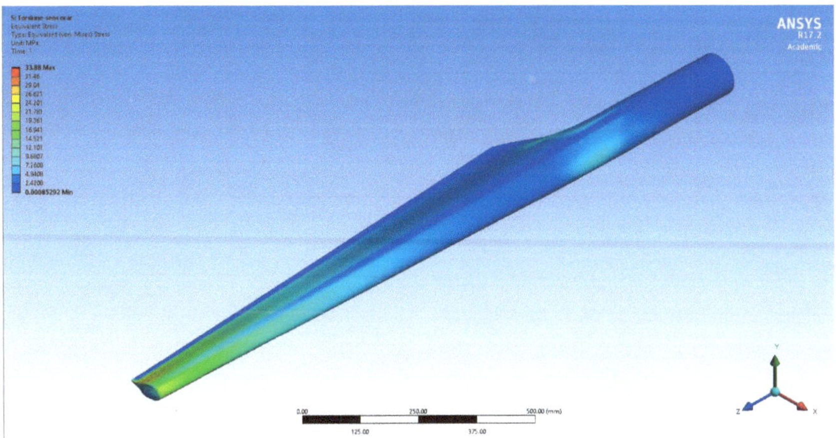

Fig 3.87 Distribution of equivalent von Mises stresses – isometric view, intrados–extrados

Figure 3.88 presents the maximum principal stress distribution, $\sigma_{1\ max}$ =32 MPa

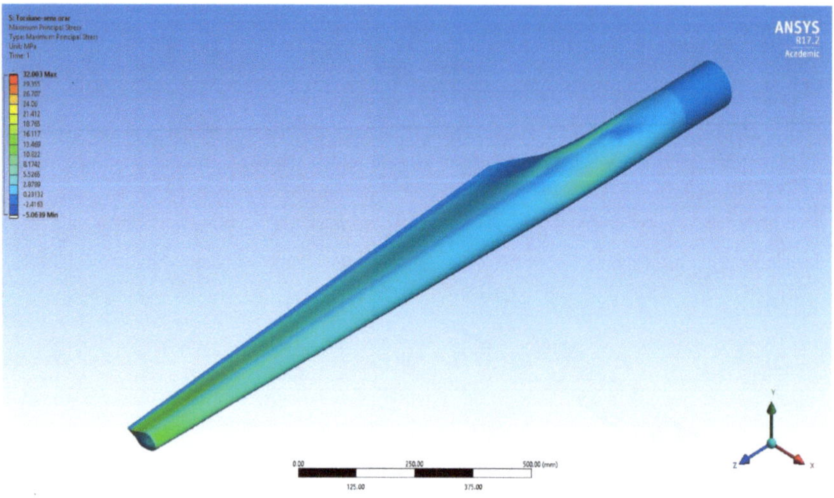

Fig 3.88 Maximum principal stress, isometric view along the intrados–extrados direction

- Torsion loading – anticlockwise direction

Figure 3.89 presents the von Mises stress distribution, $\sigma_{ech\ max}$=33.88 MPa.

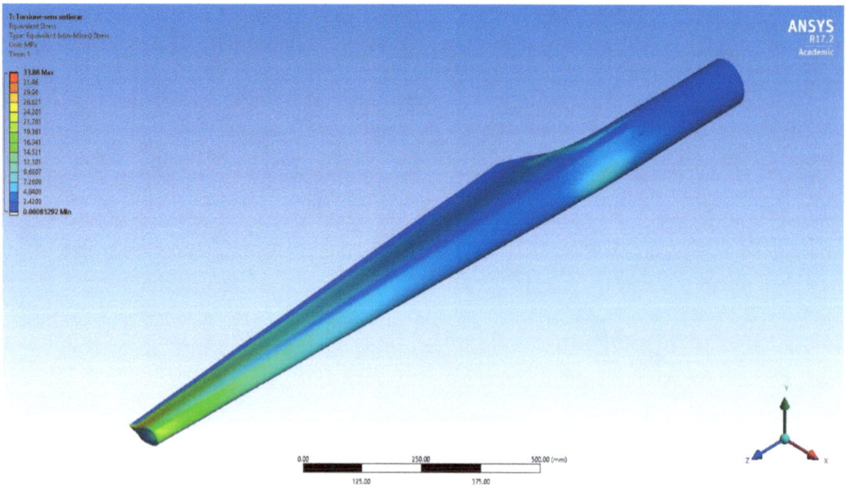

Fig 3.89 Distribution of equivalent von Mises stresses – isometric view, intrados–extrados

Figure 3.90 presents the maximum principal stress distribution, $\sigma_{1\ max}$ =31.88 MPa.

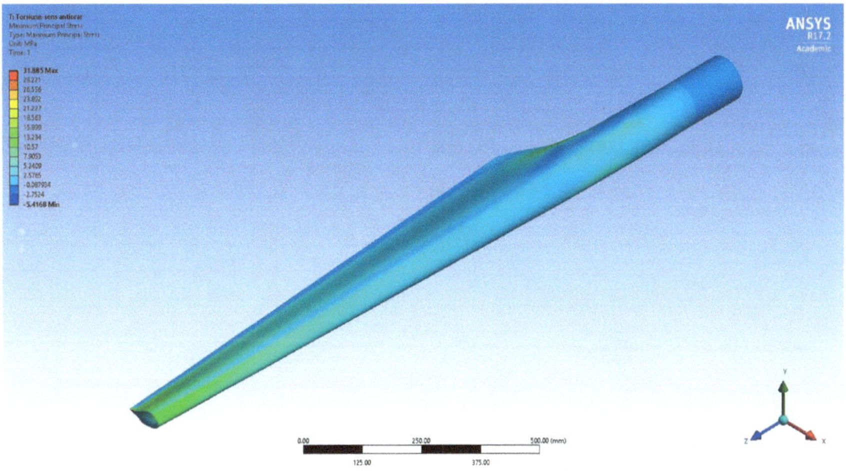

Fig 3.90 Maximum principal stress, isometric view along the intrados–extrados direction

- Bending with torsion – anticlockwise direction

Figure 3.91 presents the von Mises stress distribution, $\sigma_{ech\ max}$=47.73 MPa.

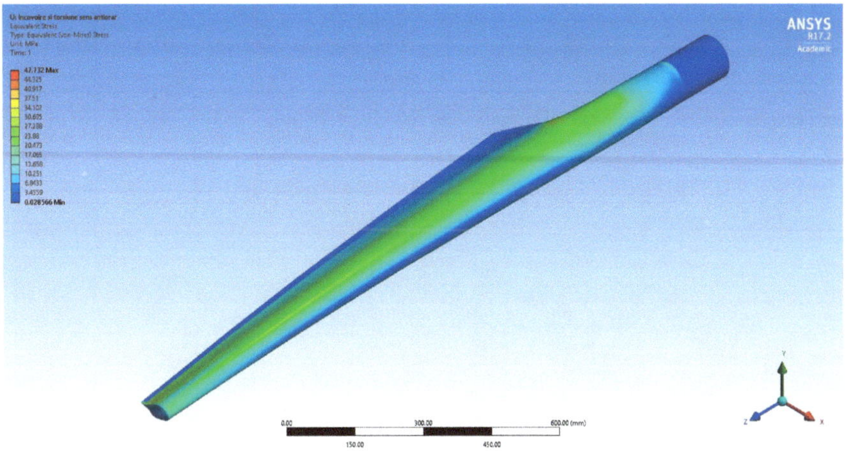

Fig 3.91 Distribution of equivalent von Mises stresses – isometric view, intrados–extrados

Figure 3.92 presents the maximum principal stress distribution, $\sigma_{1\ max}$ =34.29 MPa.

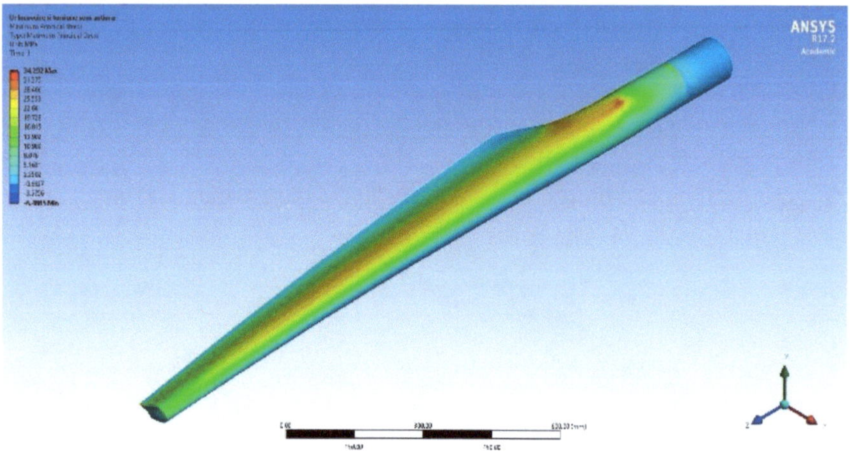

Fig 3.92 Maximum principal stress, isometric view along the intrados–extrados direction

- Bending with torsion- clockwise direction

Figure 3.93 presents the von Mises stress distribution $\sigma_{ech\ max}$=49.13 MPa.

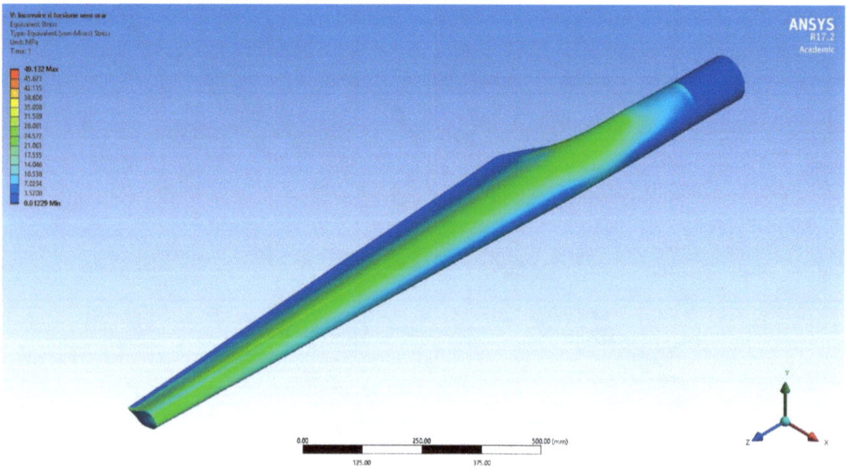

Fig 3.93 Distribution of equivalent von Mises stresses – isometric view, intrados–extrados

Figure 3.94 presents the maximum principal stress distribution, $\sigma_{1\ max}$ =39.99 MPa.

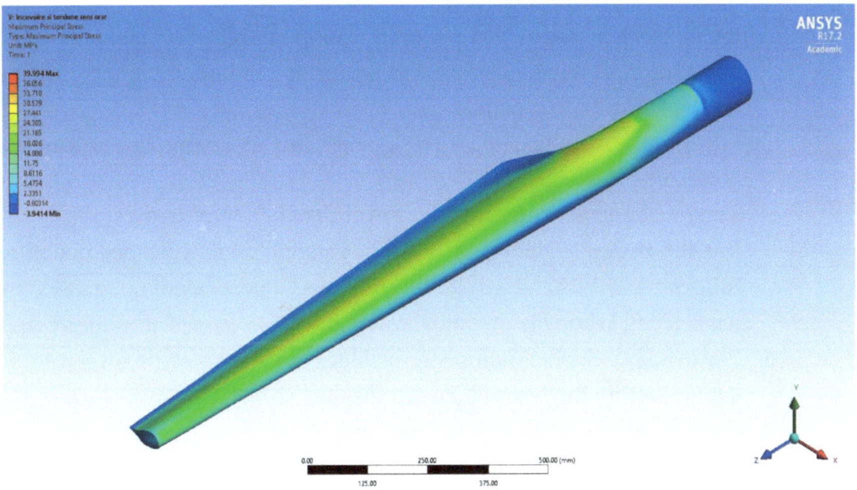

Fig 3.94 Maximum principal stress, isometric view along the intrados–extrados direction

Table 3.11 presents the relative errors obtained for the maximum von Mises stresses, determined experimentally and via FEA.

Table 3.11 The relative errors obtained for the maximum von Mises stresses

Load	σ_{exp} von Mises [MPa]	σ_{AEF} von Mises [MPa]	Errors [%]
Bending	6.43	7.09	10.26
Torsion	30.96	33.88	9.4
Compound loading	43.79	47.73	8.9

3.9 General conclusions to Chapter 3

1. Transducers 1 and 2 were positioned at a distance of 1620 mm, while transducers 3 and 4 were placed 230 mm from the point of force application. Data acquisition was performed in both directions, during both loading and unloading;
2. For simple loadings, the stress values are generally lower than for compound loadings;
3. Torsional stresses are generally lower than bending stresses;
4. For the bending loading, the strain gauge chains were positioned as follows: L1, L2, and L3 on the intrados, where bending stresses are tensile, and L4 on the extrados, where bending stresses are compressive;
5. For bending, the maximum stress values of the same type for the strain gauges within the same chain are similar;
6. For torsion, both clockwise and anticlockwise, the variations of the three stresses are similar;
7. The maximum principal stress values obtained for all strain gauges in chain 1 under the compound bending and clockwise torsion loading are higher than those obtained in the other chains;
8. The maximum principal stress values obtained for all strain gauges in chain 1 under the compound bending and anticlockwise torsion loading are higher than those obtained in the other chains.

References

1. Amenabar, I., Mendikute, A., López-Arraiza, A., Lizaranzu, M.; Aurrekoetxea, J. (2011). Comparison and analysis of non-destructive testing techniques suitable for delamination inspection in wind turbine blades. *Compos. Part B: Eng.*, 42, 1298–1305.
2. Amezquita-Sanchez, J.P., Adeli, H. (2016). Signal Processing Techniques for Vibration-Based Health Monitoring of Smart Structures. *Arch. Comput. Methods Eng.*, 23, 1–15.
3. Avdelidis, N.P., Gan, T.-H. (2013). 24- Non-destructive evaluation (NDE) of Composites: Infrared (IR) thermography of wind turbine blades. *Non-Destructive Evaluation (NDE) of Polymer Matrix Composites*; Woodhead Publishing: Sawston, UK, pp. 634–650.
4. Bang, H.-J., Jang, M., Shin, H. (2011). Structural health monitoring of wind turbines using fiber Bragg grating based sensing system. *In Sensors and Smart Structures Technologies for Civil, Mechanical, and Aerospace Systems*; SPIE: Bellingham, DC, USA, 2011; Volume 7981, pp. 716–723. https://doi.org/10.1117/12.880654
5. Baqersad, J., Poozesh, P., Niezrecki, C., Harvey, E., Yarala, R. (2014). Full Field Inspection of a Utility Scale Wind Turbine Blade Using Digital Image Correlation. CAMX 10, 2014, 2891–2960.
6. Brett, C.R., A Gunn, D., Dashwood, B.A.J., Holyoake, S.J., Wilkinson, P.B. (2018). Development of a technique for inspecting the foundations of offshore wind turbines. Insight Non-Destructive Test. Cond. Monit.60, 19–27.
7. Bulder B.H., van Dam J., Blade testing methods. Petten: Energy research Centre of the Netherlands. Report no.: ECN-C-00-055; 2002 April.
8. Carineh G., Bartosz P. and Volker T. (2021). Failure Mechanisms of GFRP Scarf Joints under Tensile Load, *Materials*, 14, 1806. https://doi.org/10.3390/ma14071806.
9. Chen C. P. and. Kam T. Y. (2011). Failure Analysis of Small Composite Sandwich Turbine Blade Subjected to Extreme Wind Load, Procedia Engineering 14 1973–1981, The Twelfth East Asia-Pacific Conference on Structural Engineering and Construction.
10. Chen X. (2017). Experimental investigation on structural collapse of a large composite wind turbine blade under combined bending and torsion. *Compos. Struct.* 160:435-445. https://doi.org/10.1016/j.compstruct.2016.10.086.

11.Chen, C., Li, H., Wang, T., Wang, L.(2020). Influence of Structural Configurations on the Shear Fatigue Damage of the Blade Trailing-Edge Adhesive Joint. *Appl. Sci.*, 10, 2715.
12.D'Amore, A., Grassia, L. (2019). Principal Features of Fatigue and Residual Strength of Composite Materials Subjected to Constant Amplitude (CA) Loading. Materials, 12, 2586.
13.Debel, C.P. (2004). Identification of damage types in wind turbine Blades tested to failure. Roskilde: Risø National Laboratory. Report no.: Risø-R-1391 (EN).
14.Dimitris Al. Katsaprakakis , Nikos P. and Ioannis N.(2021). A Comprehensive Analysis of Wind Turbine Blade Damage, *Energies*, 14, 5974. https://doi.org/10.3390/en14185974.
15.Doroshtnasir, M., Worzewski, T., Krankenhagen, R., Röllig, M. (2016). On-site inspection of potential defects in wind turbine rotor blades with thermography. *Wind Energy*, 19, 1407–1422.
16.Franko, J., Du, S., Kallweit, S., Duelberg, E., Engemann, H.(2020). Design of a Multi-Robot System for Wind Turbine Maintenance. *Energies*, 13, 2552.
17.García M., Tobias, F.P.,.Pinar Pérez A.M, Papaelias, M J.(2012). Condition monitoring of wind turbines: Techniques and methods. *Renew. Energy*, 46, 169–178.
18.Gholizadeh, S. (2016). A review of non-destructive testing methods of composite materials. *Procedia Struct. Integr*, 1, 50–57.
19.Gomez M. C.Q., García M.F.P. (2016). A new fault location approach for acoustic emission techniques in wind turbines. *Energies* 9 (1). https://doi.org/10.3390/en9010040
20.Guo, Z.-S., Zhang, J., Hu, H., Guo, X. Structural health monitoring of composite wind blades by fiber bragg grating. *In Proceedings of the International Conference on Smart Materials and Nanotechnology in Engineering, International Society for Optics and Photonics*, Harbin, China, 1–4 July 2007; p. 64230I.
21.Habibi H., Cheng L., Zheng H., Kappatos V.(2015). Selcuk C., Gan T., A dual de-icing system for wind turbine blades combining high-power ultrasonic guided waves and low-frequency forced vibrations. *Renew Energy*; 83:859–70. https://doi.org/10.1016/j.renene.2015.05.025
22.Hasen MOH (2008). Aerodynamics of Wind Turbines (Second version). UK and USA: Earthscan Press.
23.Hill, K., Meltz, G. (1997). Fiber Bragg grating technology fundamentals and overview. *J. Light. Technol.*, 15, 1263–1276.

24.https://www.mts.com/en/products/energy/wind-turbine-blade-static-systems.

25.https://www.yahoo.com/news/science-behindfrozenwindturbines212636234.html.

26.International Electrotechnical Commission.Wind turbine generator systems – part 23: full scale structural testing of rotor blades. IECTS61400-23;2001.

27.Jensen F.M., Falzon B.G., Ankersen J., Stang H.(2006). Structural testing and numerical simulation of a 34 m composite wind turbine blade, *Composite Structures* 76 52–61.

28.Jinghua W., Leian Z., Jiantao L., Xuemei H., Chengliang L. (2023). Investigation on torsional performance of wind turbine blades under pure torsion through subcomponent test, *Composite Structures* 304 116466. https://doi.org/10.1016/j.compstruct.2022.116466

29.Jinghua W., Leian Z., Jiantao Li , Xuemei H. (2023). Chengliang Li, Investigation on torsional performance of wind turbine blades under pure torsion through subcomponent test, *Composite Structures* 304 116466. https://doi.org/10.1016/j.compstruct.2022.116466

30.Jinshui Y., Chaoyi P., Jiayu X., Jingcheng Z., Suli X., Jiaotong J. (2013). Hang D., Structural investigation of composite wind turbine Blade considering structural collapse in full-scale static tests, *Composite Structures* 97 15–29. https://doi.org/10.1016/j.compstruct.2012.10.055.

31.Jørgensen E.R., Borum K.K., McGugan M., Thomsen C.L., Jensen F.M., Debel C.P., et al., Full scale testing of wind turbine Blade to failure – flapwise loading. Roskilde: Risø National Laboratory. Report no.: Risø-R-1392 (EN); 2004 June.

32.Kim, D.Y., Kim, H.-B., Jung, W.S., Lim, S., Hwang, J.-H., Park, C.-W. Visual testing system for the damaged area detection of wind power plant blade. In Proceedings of the ISR 2013, Seoul, Korea, 24–26 October 2013; pp. 1–5.

33.Kong C., Bang J., Sugiyama Y. (2005). Structural investigation of composite wind turbine blade considering various load cases and fatigue life. *Energy* 30(11– 12):2101–14. https://doi.org/10.1016/j.energy.2004.08.016

34.Lee, K., Aihara, A., Puntsagdash, G., Kawaguchi, T., Sakamoto, H., Okuma, M., (2017). Feasibility study on a strain based deflection monitoring system for wind turbine blades. *Mech. Syst. Signal Process.*, 82, 117–129. https://doi.org/10.1016/j.ymssp.2016.05.011.

35.Lei an Z., Yanzhen G., Jinghua W., Xuemei H., Xiuting W., Weisheng L. (2019). Structural failure test of a 52.5 m wind turbine blade under combined loading, *Engineering Failure Analysis* 103 286–293. https://doi.org/10.1016/j.engfailanal.2019.04.069.

36.Liu J.W., Liu P.F., Leng J.X., Wang C.Z. (2022). Finite element analysis of damage mechanisms of composite wind turbine blade by considering fluid/solid interaction. Part I: Full-scale structure, *Composite Structures* 301 116212. https://doi.org/10.1016/j.compstruct.2022.116212.

37.Liu P. F., Chen H. Y., Wu T., Liu J. W., Leng J. X., Wang C. Z. and . Jiao L. (2022). Fatigue Life Evaluation of Offshore Composite Wind Turbine Blades at Zhoushan Islands of China Using Wind Site Data, *Applied Composite Materials*, https://doi.org/10.1007/s10443-022-10098.

38.Liu P.F., Liu J.W., Wang C.Z. (2022). Finite element analysis of damage mechanisms of composite wind turbine blade by considering fluid/solid interaction. Part II: T-shape adhesive structure, *Composite Structures* 301 116211. https://doi.org/10.1016/j.compstruct.2022.116211.

39.Mads A. F., Purim L., Esben O., Lars O. B., Malcolm M. and Kim B., Structural health monitoring of 52-meter wind turbine blade: Detection of damage propagation during fatigue testing. https://doi.org/10.1017/dce.2022.20

40.Malhotra P, Hyers R W, Manwell J F, McGowan J G. (2012). A review and design study of blade testing system for utility-scale wind turbines. *Renew Sustain Energy* Rev;16(1):284–92.

41.Marín, J.C., Barrosso, A., alinís, F.; Cañas, J. (2009). Study of fatigue damage in wind turbine blades. *Engineering Failure Analysis* (16). https://doi.org/10.1016/j.engfailanal.2008.02.005.

42.Márquez, F.P.G., Chacón, A.M.P. (2020). A review of non-destructive testing on wind turbines blades. *Renew. Energy*, 161, 998–1010. https://doi.org/10.1016/j.renene.2020.07.145.

43.Morăras C., Goantă V., Husaru D., Istrate B., Paul Doru Barsanescu and Corneliu Munteanu.(2023). Analysis of the Effect of Fiber Orientation on Mechanical and Elastic Characteristics at Axial Stresses of GFRP Used in Wind Turbine Blades, *Polymers*, 15(4), 861; https://www.mdpi.com/2073-4360/15/4/861.

44.Morăras C., Goantă V., Istrate B., Munteanu C. Dobrescu G. (2022). Structural Testing by Torsion of Scalable Wind Turbine Blades. *Polymers*, 14, 3937. https://doi.org/10.3390/polym14193937.

45.Morăraş, C., Barsanescu, P D., Tugui, C A. (2020). Some aspects regarding the influence of humidity on the tensile characteristics of GFRP composite material, *IOP Conference. Series: Materials Science and Engineering* 997doi:10.1088/1757-899X/997/1/012100.

46.Ozbek, M., Rixen, D.J. (2013). Operational modal analysis of a 2.5 MW wind turbine using optical measurement techniques and strain gauges. *Wind Energy*, 16, 367–381. https://doi.org/10.1002/we.1493.
47.Qiao,W., Lu, D. (2015). A Survey on Wind Turbine Condition Monitoring and Fault Diagnosis—Part II: Signals and Signal Processing Methods. *IEEE Trans. Ind. Electron*. 62, 6546–6557. DOI: 10.1109/TIE.2015.2422394
48.Rocha I.B.C.M, (2017). Hygrothermal ageing behaviour of a glass/epoxy composite used in wind turbine blades, Compos. Struct. 174 110–122.
49.Rocha I.B.C.M., et al., (2017). Hygrothermal ageing behaviour of a glass/epoxy composite used in wind turbine blades, *Compos. Struct*. 174 110–122. https://doi.org/10.1016/j.compstruct.2017.04.028.
50.Rodrigo P.T., Vincent B., W. Van Paepegem. (2022). Finite element analysis of wind turbine blades subjected to torsional loads: Shell vs solid elements, *Composite Structures* 280 114905. https://doi.org/10.1016/j.compstruct.2021.114905.
51.Sanati, H., Wood, D., Sun, Q., Condition Monitoring of Wind Turbine Blades Using Active and Passive Thermography (2018). *Appl. Sci*., 8, 2004. https://doi.org/10.3390/app8102004
52.Savin A., Iftimie N., Steigmann R., Rosu D., Dobrescu G. (2018). Effective Methods for Structural Health Monitoring of Critical Zones of Scalable Wind Turbine Blades. doi:10.5545/sv-jme.2017.5130.
53.Schroeder, K., Ecke,W., Apitz, J., Lembke, E., Lenschow, G.(2005). A fibre Bragg grating sensor system monitors operational load in a wind turbine rotor blade. *Meas. Sci. Technol*, 17, 1167–1172. DOI 10.1088/0957-0233/17/5/S39.
54.Sørensen B.F., Jørgensen E.R., Debel C.P., Jensen F.M., Jensen H.M., Jacobsen T.K., et al, Improved design of large wind turbine Blade of fiber composites based on studies of scale effects (Phase1) – summary report. Roskilde: Risø National Laboratory. Report no.: Risø-R-1390 (EN); 2004 September.
55.Tang, J., Soua, S., Mares, C., Gan, T.-H. (2016). An experimental study of acoustic emission methodology for in service condition monitoring of wind turbine blades. *Renew. Energy*, 99, 170–179.
56.Tarfaoui M, Shah OR, Nachtane M. (2019). Design and optimization of composite offshore wind turbine blades. *J Energ Resour Technol*;141(5):051204.
57.Tiwari, K.A., Raisutis, R. (2018). Post-processing of ultrasonic signals for the analysis of defects in wind turbine blade using guided waves. *J. Strain Anal. Eng. Des*., 53, 546–555.

58. Wang, Y., Yoshihashi, R., Kawakami, R., You, S., Harano, T., Ito, M., Komagome, K., Iida, M., Naemura, T. (2019). Unsupervised anomaly detection with compact deep features for wind turbine blade images taken by a drone. *IPSJ Trans. Comput. Vis. Appl.*, 11, 3.
59. William F., Yadong J., Nicolas D., Peter D. and Jamie G. (2021). Investigation and Validation of Numerical Models for Composite Wind Turbine Blades, *J. Mar. Sci. Eng.*, 9, 525. https://doi.org/10.3390/jmse9050525.
60. Wu, J., Song, C.; Saleem, H.S.; Downey, A.; Laflamme, S. (2015). Network of flexible capacitive strain gauges for the reconstruction of surface strain. *Meas. Sci. Technol.*, 26, 055103.
61. www.micro-measurements.com: General Purpose Strain Gages—Stacked Rosette, 062WW.
62. Xiao C. (2017). Experimental investigation on structural collapse of a large composite wind turbine blade under combined bending and torsion, *Composite Structures* 160 435–445. https://doi.org/10.1016/j.compstruct.2016.10.086.
63. Xiao C. (2019). Experimental observation of fatigue degradation in a composite wind turbine blade, *Composite Structures* 212 547–551. https://doi.org/10.1016/j.compstruct.2019.01.051.
64. Xiao C., Wei Z., Xiao Lu Zhao and Jian Zhong Xu. (2014). Failure Test and Finite Element Simulation of a Large Wind Turbine Composite Blade under Static Loading, *Energies*, 7, 2274-2297; https://doi.org/10.3390/en7042274.
65. Yadong J., William F., Finlay W., Michael F., Tomas F., Jamie G. (2023). Structural analysis of a fibre-reinforced composite blade for a 1 MW tidal turbine rotor under degradation of seawater, *Journal of Ocean Engineering and Marine Energy*, doi.org/10.1007/s40722-023-00279-w.
66. Yang, J., Peng, C., Xiao, J., Zeng, J., Yuan, Y. (2012). Application of videometric technique to deformation measurement for large-scale composite wind turbine blade. *Appl. Energy*, 98, 292–300.
67. Zhou H.F., Dou H.Y., Qin L.Z., Chen Y., Ni Y.Q., Ko J.M. (2014). A review of full-scale structural testing of wind turbine Blades, *Renewable and Sustainable Energy Reviews* 33177–187. https://doi.org/10.1016/j.rser.2014.01.087.
68. Zhou, W., Li, Y., Li, Z., Liang, X., Pang, Y., Wang, F. (2014). Interlaminar Shear Properties and Acoustic Emission Monitoring of the Delaminated Composites for Wind Turbine Blades. In Advances in Acoustic Emission Technology; Springer: Berlin/Heidelberg, Germany; pp. 557–566.

www.ingramcontent.com/pod-product-compliance
Lightning Source LLC
LaVergne TN
LVHW051128210926
848482LV00001B/15